George Rantoul White

An elementary chemistry

George Rantoul White

An elementary chemistry

ISBN/EAN: 9783337275808

Printed in Europe, USA, Canada, Australia, Japan

Cover: Foto ©berggeist007 / pixelio.de

More available books at **www.hansebooks.com**

AN

ELEMENTARY CHEMISTRY

BY

GEORGE RANTOUL WHITE, A.M.

INSTRUCTOR IN CHEMISTRY AT PHILLIPS EXETER ACADEMY

BOSTON, U.S.A.
PUBLISHED BY GINN & COMPANY
1894

Copyright, 1894
By GEORGE RANTOUL WHITE

ALL RIGHTS RESERVED

PREFACE.

THIS book is little more than a reproduction of the course in elementary chemistry as now given at Exeter Academy. The course itself has been developed, little by little, during several years of observation and experiment on the part of the writer, to meet the needs of all classes of students, — those who are preparing for a further course of study at college, those who expect to enter a scientific school, and those who go from the academy directly to their life-work. The majority of all these students take chemistry merely as a part of a liberal education, some intend to follow the paths of science; a few will become chemists.

In planning this work for beginners the writer has tried to prepare a course that will meet the needs of one class as well as those of another. But in this respect his task has been easy, for the more he has considered the needs of the various classes, the more he has come to believe that the elementary training of all should be alike. The student who is to be a lawyer, a doctor, or a man of business, needs that same careful attention to details, that same power of accurate observation which is expected of the coming chemist; and he who is to be the chemist needs the same high development of his reasoning powers as he who takes chemistry only for the intellectual training it can give.

PREFACE.

This book is designed especially for the use of two classes of students. First for those whose instruction is placed in the hands of a teacher who cannot devote his whole time to chemistry; and secondly, for young men and women who are eager to study chemistry but have no teacher at all.

In regard to the first class — those instructed by teachers who are not strictly and solely teachers of chemistry — the writer believes that it is not possible for any teacher to get the best results from his students if he has to divide his attention, his energy, his love, between two or more subjects. We cannot serve two masters. Those students who are so fortunate as to have an instructor who can devote his whole time to the presentation of elementary chemistry need no book at all. The instructor will himself more than take the place of a book. But it is seldom that a school or college has a chemist who can devote his whole time to developing the elementary course. Generally the instructor in chemistry must teach some other study. Even when he can give all his attention to chemistry, his time often is chiefly devoted to the more advanced students. Though the instructor can well replace the book, no book can fully take the place of an instructor. It is hoped, however, that this book may be of some service in supplementing the work of those instructors who have not the time to do for their pupils all that they really desire.

To those students who can never have any instructor, but want to study chemistry, the strictly inductive method here followed lends itself well. There are, moreover, no experiments inserted that have not been well tried, and found to work successfully in the hands of beginners. Wherever there seems danger — and danger there must be, to some extent, in every course of chemistry that is worth the giving — abundant warning is inserted. Through all the development of this course at Exeter there has never been an accident at all serious.

It will be noted by him who looks through the pages of this book, first, that the method of presenting elementary chemistry here embodied is preëminently a practical or laboratory method, and, secondly, that the method is preëminently inductive.

To-day no plea for the laboratory or practical method of presenting any study seems needed. As James Russell Lowell has said, "Practical application is the only mordant which will set things in the memory. Study without it is gymnastics, and not work, which alone will get intellectual bread." In no branch of learning does the laboratory method seem more essential than in chemistry, and in no branch has this method been more widely adopted than in chemistry. Everywhere laboratories are being built, and nowhere, so far as the writer has ever heard, has there been a change back to the old method after the new has once been tried. The results, however, always justify the change from the old to the new. Professor Cooke, director of the Chemical Laboratory at Harvard University, where for years chemistry has been taught in no other way than by the laboratory method, writes, "How little of what we value comes to us by the way of direct teaching! There may be such a thing as too much teaching, and even — in spite of the paradox — too good teaching; for, after all, personal experience is, at the university as elsewhere, the most efficient teacher, and he who encourages us to help ourselves is our safest guide."

In working out the course indicated in this book it has been the writer's aim to make this "personal experience" for the student as large as possible. He always tells his Exeter students that he is their companion more than their teacher; that their own experience, acquired little by little, through hours of work in the laboratory, must be their true teacher; while he himself simply is their guide, showing where the path runs smoothest, warning against innumerable by-paths, pleasant to follow it is true, but that lead nowhere, pointing out the dangers to be avoided and, alas! not infrequently having to point out the beauties that other travelers on the same way have noted, — beauties that would otherwise be passed by, all unheeded by the impetuosity of youth.

At first the student is told little or nothing. He is compelled to find out all things for himself. To assist him in finding the essential, and to make sure that he has succeeded in this, frequent questions are inserted in the text of the experimental part. [For the use to be made of these questions, see Introduction.] The author believes that questions inserted thus, just at the point when the inquiry naturally springs to the lips of a bright student, are of more value than the same questions inserted all together at the end of a book or even in groups at the end of the chapters.

There is a marked tendency among bright students who are taking up a new subject for study, to ask questions. This tendency can readily be checked by the teacher's refusing to answer such questions and telling the pupils to confine themselves to the text of the book; or it can be developed, by proper training, into one of the highest and most valuable powers of the human mind — logical reasoning. It was by attempting to win something of good from this asking of questions that the author of this book was led to arrange the work of his elementary class on a strictly inductive basis. A fresh young mind is very loath to take anything for granted, — it is naturally questioning. To be sure, it is always easy to destroy this natural condition, and to accustom a pupil to have implicit faith in his teacher; implicit faith in his book; in fact, from the very method by which we have most of us been taught, we are obliged to say, if questioned as to our reason for believing a certain statement, that we believe it because so and so has told us, or because such and such a book says so. How much more of profit and pleasure will come to a future generation if only we can teach the children of to-day — who are to be the men and women of to-morrow — to think and to reason for themselves. As President Eliot of Harvard has recently said: "The main processes or operations of the mind" which should be developed "in an individual in order to increase his general intelligence and train his reasoning power" are, first, "*observation*, that is to say, the alert, intent, and accurate use of his senses"; next, "the function of making a correct record of things observed"; next, "the faculty of drawing correct inferences from recorded observations. . . . It is often a long way from the patent fact to the just inference. For centuries the Phoenician and Roman navigators had seen the hulls of vessels disappearing below the blue horizon of the Mediterranean while their sails were visible; but they never drew the inference that the earth was round."

It is easier to learn [?] a thing from a book or from a teacher than by reasoning it out for ourselves. But do we get from the memorizing process the results that are needed for an education? There must come a time when almost every student will find himself in a great school where he himself is not pupil but master, where there are no books to which he can refer. Herbert Spencer, speaking of the state of education, has said: "Nearly every subject dealt with is arranged in abnormal order; definitions, and rules, and principles being put first, instead of being disclosed, as they are in the order of nature, through the study of cases. And then pervading the whole is the vicious system of rote learning — a system of sacrificing the spirit to the letter. See the results." Though there have been marked changes since these words were written, some thirty years ago, still there is at the present day much "rote learning," even among chemical students.

As an instructor for several summers in the Harvard Summer School at Cambridge [where teachers from almost all parts of the country assemble], the writer has had an opportunity to judge of the methods of instruction generally employed, and to "see the results." To him the results have not seemed satisfactory. Those taught by the common methods, at the best, have a considerable knowledge of facts, and sometimes can make a few simple analyses, but *they seldom have the power of doing any thinking for themselves.* Believing, as the author does, that there is no other study which, pursued for either one, two, or three years, can give so much in return that is of an educational value, he has been grieved and ashamed to realize how little *is* obtained by the many, mostly faithful, students of that science. Seeking for the cause, the writer has come to the conclusion that it lies in the method of presenting the subject. It is seldom that a pupil knows much of anything about this science till he has studied it four, five, or more years. Then the real meaning of all he has been working over gradually dawns upon him. But how is it with the great majority of one, two, and three year students who, never intending to be chemists, have taken up the study only as a part of a liberal education? Most of these have actually wasted the time spent on their chemistry. The trouble is that most text-books aim at giving a synopsis of the whole of chemistry. The writers do not seem to realize how vast an accumulation of facts now lie open to the chemist: that to try to learn all these facts is as profitless a task as for the beginner in English composition to start by learning all the words in his dictionary. Is it not better for both to confine themselves to a little that is of everyday occurrence and learn that little thoroughly? Then again, to the writer, it seems that the text-books for beginners err in being "arranged in abnormal order: definitions, and rules and principles being put first, instead of being disclosed, as they are in the order of nature, through the study of cases." In the following pages a "study of cases" comes first. In fact for some time the pupil does nothing but study cases; very simple ones at first — many, though, that have been familiar to him all his life; then less familiar and more complicated ones; and, finally, in Ex. 32, "A Chemical Investigation," he has before him a problem not unlike the many that are constantly presented to the chemist who is working on the borderland of the science. To the end of this "Chemical Investigation," the whole course is inductive. It is believed that not an essential question can be asked that the student cannot answer either from observation of the phenomena or from reasoning based on previous experiment. Through the whole of Part I the student becomes acquainted with the methods — both mental and practical — of the scientific chemist of to-day. He learns *to experiment, to observe, to reason.* He accepts nothing simply because his teacher, or the book, says it is so. Everything he verifies by himself to his own satisfaction. Then, having learned by many experiments [particularly by "A Chemical Investigation"] how the chemist obtains the facts that are at the basis of all his reasoning, the student is prepared to trace with pleasure [in Part III] the history of chemistry, to note what observations lead to the establishment of certain theories, and the recognition of what facts lead to the overthrow of these same theories; to recognize the gradual unfolding of chemical law; and, finally, to inspect the foundations on which our present Atomic

Theory rests, and have an opinion of his own as to its stability. The author has thought best to give a more extended account of the development of the laws and theories of chemistry than is usually found in text-books. He has been led to this for several reasons: first, because he believes that every educated man or woman should possess a knowledge of these laws and theories; secondly, because he feels that a consideration of their development gives the student an excellent introduction to the true spirit of scientific investigation; and, finally, because he is well aware that advanced students of chemistry generally lack a fundamental knowledge of the history and development of their own branch of science. Text-books for beginners usually omit a thorough treatment of the laws and theories as beyond their scope, while the books for advanced students also omit the same on the ground that the student has already learned the fundamental principles. Where, then, is the student to get this knowledge?

It may also be noted that for a long time no hint is given that there are such things as chemical symbols. Not to accustom the pupils from the first to express facts in the language of chemistry, will doubtless seem to many both silly and wrong. But to do so would not be in accord with the spirit of the inductive method here employed. When the student has mastered the facts, it is a pleasure to see how readily he expresses those facts by the proper symbols. This he then does accurately and almost without an effort. The writer several years ago adopted the plan of having his students express the facts in English till enough facts had been collected for the students to use the language of chemistry freely and accurately. He has no desire to go back to the old method. Then, too, when symbols are used before they can be fully understood, there is danger that the symbols, and not the facts they express, may be looked at as the realities. For instance, on a recent [English] examination paper there appeared the following: "Account for the basicity of phosphorus and hypophosphorus acids, respectively, by reference to their constitutional formulae." At the best, formulae can only express the facts we have found out about substances; they cannot *account for* any of these facts, and the student should not be taught that they can. In this connection it may be of interest to note the following taken from the introduction to the first edition of Meyer's "Moderne Theorien der Chemie." The translation reads, "Chemical symbols and formulae, which a few years ago received such prominence, are now regarded with indifference, since what was formerly expressed symbolically and indistinctly or even without proof or clearness by their aid, can now be expressed in clear words with fixed meaning." And again, very recently, no less an authority than Professor Remsen, of Johns Hopkins University, has written that he "sometimes thinks, and the intervals between the thoughts are getting shorter, that if the use of formulae were given up entirely in elementary instruction, better results would be obtained."

February, 1894.

CONTENTS.

	PAGE
INTRODUCTION	xxiii

PRELIMINARY WORK

Measuring	3
Weighing	4
Making a wash-bottle	6

PART I. EXPERIMENTS.

Ex. 1. Iron
    A. Properties ... 11
    B. With Air ... 11
2. Phosphorus
    A. Properties ... 12
    B. With Air ... 13
3. Mercury ... 14
4. Carbon ... 14
5. Artificial Preparation of Oxygen
    A. From Red Oxide of Mercury ... 15
    B. From Chlorate of Potassium ... 16
6. Action of Undiluted Oxygen Gas
    A. On Iron ... 17
    B. On Phosphorus ... 18
    C. On Carbon ... 18
7. Preparation of Oxygen from Water ... 19
8. Examination of that Constituent of Water which is not Oxygen ... 20
9. Hydrogen
    A. Preparation from Water by means of Iron ... 20
    B. Specific Gravity ... 22
    C. Action of Air on Warm Hydrogen ... 22
10. Sulphur
    A. Properties ... 23
    B. Modifications ... 24
    C. Action of Oxygen on Hot Sulphur ... 25

CONTENTS.

	PAGE
Ex. 11. Sulphurous Acid	27
12. A Second Oxide of Sulphur	27
13. Sulphuric Acid	29
14. Removal of Hydrogen from Sulphuric Acid	30
15. Action of Water on Oxide of Iron	32
16. Action of Water on Oxide of Phosphorus	33
17. Action of Water on Oxide of Carbon	33

18. Zinc
  - A. Properties 34
  - B. Oxidation of Zinc 34
  - C. Action of Water on Oxide of Zinc 34
19. Action of Zinc on the Non-combustible Oxide of Carbon; or,
  Preparation of a Second Oxide of Carbon 35
20. Oxidation of the Combustible Oxide of Carbon 37
21. Action of Zinc on Sulphuric Acid 37
22. Action of Oxide of Zinc on Sulphuric Acid 38
23. Sulphides
  - A. Mutual Action of Iron and Sulphur 39
  - B. Action of Hydrogen on Warm Sulphur 39
  - C. Action of Sulphuric Acid on Sulphide of Iron.. 40
24. Copper
  - A. Properties 41
  - B. Oxidation 41
  - C. Reduction of the Black Oxide 42
25. Magnesium
  - A. Properties 43
  - B. Oxidation 43
  - C. Magnesium Oxide with Water 43
  - D. Magnesium and Sulphuric Acid 43
26. Calcium
  - A. Properties 44
  - B. Oxidation 44
  - C. Reaction of Oxide of Calcium and Water 45
  - D. Action of Calcium on Water 45
  - E. Reaction of Hydrate of Calcium and Sulphuric Acid 46
  - F. Reaction of Hydrate of Calcium and Carbonic Acid 47

CONTENTS. XV

	PAGE
Ex. 26. Calcium	
G. Analysis of Marble	50
H. Reaction of Marble and Sulphuric Acid	51
27. Sodium	
A. Properties	52
B. Oxidation	52
C. Reaction of Oxide of Sodium and Water	53
D. Action of Sodium on Water	53
E. Reaction of Sodium Hydroxide and Sulphuric Acid	54
F. Reaction of Sodium Hydroxide and Carbonic Acid	54
G. Reaction of Hydrate of Sodium and the Non-combustible Oxide of Carbon	55
H. Reaction of Carbonate of Sodium and Sulphuric Acid	55
I. Reaction of Sodium and Mercury	55
28. Chlorine	56
29. Chlorides	
A. Chloride of hydrogen	56
B. Chloride of Sodium	57
C. Preparation of Hydrochloric Acid on a Large Scale	57
D. Solubility of Hydrochloric Acid	58
E. Reaction of Hydrochloric Acid and Marble	58
F. Action of Sodium on Hydrochloric Acid	59
G. Action of Sodium Hydroxide with Hydrochloric Acid	60
30. Potassium	
A. Properties	61
B. Oxidation	61
C. Reaction of the Oxide and Water	61
D. Reaction of Potassium and Water	61
E. Action of Potassium on the Dioxide of Carbon	61
F. Reaction of Potassium Hydroxide and Sulphuric Acid	62
31. Nitrogen	63
32. A Chemical Investigation	63
A. Preparation of Nitric Acid	64
B. Action of Magnesium on Nitric Acid	65

CONTENTS.

	PAGE
Ex. 32. A Chemical Investigation	
C. Action of Copper on Nitric Acid	65
D. Action of Carbon on Nitric Acid	66
E. Reaction of Nitric Acid and Potassium Hydroxide	67
33. Ammonia	
A. Preparation of Ammonia	68
B. Ammonia Fountain	70
C. Salts of Ammonia	71
34. Oxides of Nitrogen	72

PART II. ADDITIONAL EXPERIMENTS.

Ex. 1. Bromine	77
2. Bromides	78
A. Properties of Hydrogen Bromide	78
B. Sodium Bromide	78
C. Replacement of Bromine	78
3. Iodine	
A. Properties	79
B. Solubility	79
C. Action on the Skin	79
D. Action on Starch	79
4. Iodides	
A. Properties of Potassic Iodide	80
B. Replacement of Iodine by Chlorine	80
C. Will Bromine Displace Iodine?	81
5. Fluorine and Fluorides	
A. Properties of Calcic Fluoride	81
B. Preparation of Fluoride of Hydrogen	81
C. Etching of Glass by Hydrofluoric Acid	82
6. Arsenic and its Compounds	
A. Properties of Arsenic	83
B. Oxidation	83
C. Reduction of the Oxide of Arsenic	84
D. Arsenide of Hydrogen	84
E. Detection of Arsenic	85
7. Antimony	
A. Properties	87
B. Oxidation	87
C. Chloride	88

		PAGE
Ex. 7.	Antimony	
	D. Hydrogen Antimonide	88
	E. A Chemical Examination	89
8.	Bismuth	
	A. Properties	89
	B. Nitrate of Bismuth	89
9.	Tin	
	A. Properties	89
	B. Oxidation	90
	C. Crystalline Structure	90
	D. Action of Strong Acids with Tin	90
	E. Replacement of Tin by Zinc	91
10.	Lead	
	A. Properties	91
	B. Oxidation	91
	C. Action of Water on Oxide of Lead	91
	D. Action of Acids on Lead	91
	E. Replacement of Lead by Zinc	92
	F. Lead Chloride	92
	G. Lead Sulphate	92
	H. Plumbers' Solder	92
	I. Fusible Alloy	93
11.	Silver	
	A. Properties	93
	B. Oxidation	93
	C. Action of Acids on Silver	93
	D. Replacement of Silver by Copper	94
	E. Replacement of Silver by Calcium, Sodium and Potassium	94
	F. Sulphide of Silver	95
	G. Oxide of Silver	95
	H. Purification of Silver	95
12.	Gold	
	A. Properties	96
	B. Action of Acids on Gold	96
	C. Chloride of Gold	97
	D. Gold Amalgam	97
	E. Color of Gold	97
13.	Platinum	
	A. Properties	98

CONTENTS.

		PAGE
Ex. 13. Platinum		
B. Action of Acids on Platinum		98
C. Action of other Chemicals, besides Acids, on Platinum		98
D. Action of Metals with Platinum		98
E. Platinum Sponge		98
14. Aluminum		
A. Properties		99
B. Oxidation		99
C. Action of Acids on Aluminum		99
D. Sulphate of Aluminum		99
E. Alum		100

PART III. HISTORY AND DEVELOPMENT OF THE LAWS AND THEORIES OF CHEMISTRY.

CHAPTER I. INTRODUCTION ... 103
  Physical and Chemical Changes ... 103
    Ex. 1. Two kinds of Changes ... 104
       2. Changes caused by Water of Crystallization 106
       3. Change caused by the Action of Sulphuric Acid on Water ... 106
  Analyses, Syntheses, and Metatheses ... 107
    Ex. 4. Synthesis of Chloride of Ammonium ... 108
       5. Metatheses ... 110

CHAPTER II. THE EARLIEST PERIOD ... 111

CHAPTER III. THE PERIOD OF ALCHEMY ... 114
    Ex. 6. A So-called Transmutation ... 116
       7. Death of a Metal ... 117
       8. Resurrection of a Metal ... 117

CHAPTER IV. THE MEDICAL PERIOD ... 120

CHAPTER V. THE PERIOD OF ROBERT BOYLE ... 125
    Ex. 9. The Law of Boyle ... 126
      10. Qualitative Tests ... 132
        A. Tests used by Boyle ... 132
        B. Tests by Physical Changes ... 132
        C. Tests by Chemical Changes ... 133
      11. Mechanical Mixture and Chemical Compound ... 135
        A. Iron and Sulphur ... 135
        B. Zinc and Sulphur ... 136

CONTENTS.

	PAGE
CHAPTER VI. THE PHLOGISTON PERIOD	138
CHAPTER VII. THE PNEUMATIC PERIOD	140
Ex. 12. Weight and Specific Gravity of Air	141
13. The Law of Dalton	143
14. Weight and Specific Gravity of Carbonic Dioxide	145
15. Weight and Specific Gravity of Hydrogen Gas	148
16. Weight and Specific Gravity of Illuminating Gas	149
17. Conservation of Mass	157
A. The Combustion Products of a Candle	157
B. The Weight of the Products is Equal to the Weight of the Factors	159
CHAPTER VIII. THE MODERN, OR ATOMIC THEORY, PERIOD	161
§ 1. Introductory	161
Ex. 18. Law of Definite Proportions by Weight	163
§ 2. Quantitative Analysis	164
Ex. 19. Analysis of Table Salt	165
§ 3. Multiple Proportions	168
Ex. 20. Multiple Proportions	169
A. The Oxides of Sulphur	169
B. The Oxides of Nitrogen	169
C. The Chlorides of Iron	170
§ 4. Dalton's Atomic Theory	170
§ 5. Combining Number	172
Ex. 21. Determination of the Combining Number for Zinc	172
§ 6. Prout's Hypothesis	175
§ 7. Molecules	176
§ 8. Relative Weight of the Atoms	177
Ex. 22. Law of Definite Proportions by Volume	179
§ 9. The Molecular Theory	180
Ex. 23. Spaces between the Molecules	182
24. Irregular Expansion of Liquids	184
25. Regular Expansion of Gases	185
§ 10. Determining Atomic Weights	192

CONTENTS.

	PAGE
§ 11. Determining Molecular Weights	195
Ex. 26. Determination of Molecular Weights by the Physical Method	196
A. Molecular Weight of Carbonic Dioxide	196
B. Molecular Weight of Oxygen Gas	196
27. Determination of Molecular Weights by the Chemical Method	197
A. Molecular Weight of Chlorate of Potassium	198
B. Molecular Weight of Chloride of Potassium	199
C. Molecular Weight of Sulphate of Potassium	200
§ 12. Specific heat	202
Heat	203
Ex. 28. Transference of Motion	203
29. Specific Heat of Zinc	205
30. Specific Heat of Iron	207
31. Specific Heat	207
Law of Dulong and Petit	208–209
Determination of the True Atomic Weight of Zinc from its Combining Number and its Specific Heat	210
§ 13. Isomorphism	211
Ex. 32. Isomorphism	211
§ 14. The Periodic Law	213
Table of Atomic Weights	216
LANGUAGE OF CHEMISTRY	218
Table of Atomic Symbols	219
STOICHIOMETRY	224
MANIPULATIONS	226
To Mark Glass	226
To Cut Glass	226
To Fire-Polish the Edges of Glassware	228
To Bend Glass Tubes and Rods	228
To Draw out Glass Tubes	229
To Make a Matrass	230
To Render Corks Air-Tight	230

CONTENTS.

MANIFULATIONS   PAGE

   To Render Joints Air-Tight 231
   To Cut Rubber Neatly and Quickly 231
   To Pass a Glass Tube through a Hole in a Rubber Stopper 231
   To Bore a Round Hole in Glass 231
   To Prevent Mixing Glass Stoppers 232
   To Hold Hot Beakers, Test-Tubes, etc. 232
   To Use the Pneumatic Trough 232
   To Use Filter Papers 233
   To Dry Bottles, Flasks, etc. 234
   To Remove Stoppers that have Stuck 234
   To Pour Gases 235
   To Use a Bunsen Burner 235
   To Use the Bunsen Blast-Lamp 236

APPENDICES 237

   A. Apparatus for the Electrolytic Decomposition of Water .. 239
   B. Hydrogen Explosions 241
   C. Test Papers 242
   D. Suction Pumps 243
   E. Catch-Bottles 244
   F. Generator for Gases 246
   G. Hood 247
   H. Preparation of Chlorine 247
   I. Sodium Amalgam 248
   J. Test Solutions 249
   K. Use of the Mouth Blow-Pipe 250
   L. Arsenical and Antimonial Papers for Testing 250
   M. To Dry Precipitates 251
   N. To Nurse a Crystal 251
   O. Distilled Water 252
   P. Directions for a Student Who Has no Instructor 254

INDEX 259

INTRODUCTION.

THE student should provide himself with a blank-book — best, one with no ruling whatever — with pages not less than six by eight inches, and containing not less than 150 *leaves;* [1] also an apron large enough to cover chest as well as legs, or, better, overalls and a light, cheap workingman's jacket. Before the time appointed for the first work in the laboratory he should apply to his instructor for the assignment of desk, apparatus, and chemicals, and the directions for beginning work.[2]

The first three exercises are preliminary to the regular work. Their chief purpose is to enable the pupil to form the acquaintance of the system of weights and measures used in scientific work, and to accustom himself to the use of apparatus.

In all scientific work it must be the student's aim to *observe*, to *experiment*, to *reason*. His goal is the Truth. Let him continually ask the question "Why?" The

---

[1] Experience has shown that it is best to get a blank-book with leather back and tips. Such a book should not cost over thirty-five or forty cents. Laboratory usage is apt to destroy the light cloth back of the ordinary pasteboard-covered blank-book. The common blank-book also quickly wears out at the corners, and, as the leather tips cost but little, they are recommended.

[2] A student who is going to use this book without an instructor will find his directions in Appendix P.

laboratory note-book must be a store-house for the observations from the experiments. A full page should be reserved before beginning the description of any experiment [or before that of every part of an experiment when the parts themselves are long or promise to demand much arithmetical calculation]. On this page the student should put his first rough notes *at the moment the observations are made.* Here also should appear all figuring, the results of which only he wishes to appear in the course of his description. There must be *no note-taking nor figuring of any kind on bits of paper, apparatus, etc.* All the notes, and the whole figuring for mathematical problems must appear. Remember that a page is particularly reserved for this rough work, and the student must not allow himself to form the habit of entrusting his notes to papers, that are always apt to be misplaced or destroyed. Moreover, if a mistake has been made in number only, it is irritating, indeed, to be obliged to repeat all the experiment when, if the original data were present, the mistake could be corrected, often at a glance. Do not correct mistakes by erasure and rewriting. Cross out the old and let the new appear by its side. Do not fear that this method will spoil the appearance of the book in the eyes of the instructor. Every instructor of experience knows that, though greatly to be sought after, infallibility is not attainable by a beginner in laboratory work, or, in fact, by any other student, while erasures, even by the best of students, must be looked at with some suspicion. Try, however, to confine all correction of mistakes to these pages just mentioned as reserved for rough notes and figuring. Do not attempt to write the fair account of the experiment till the experiment has been performed completely and satisfactorily in the laboratory and the necessary observations and calculations [if any] have been made on the preliminary page. When the experiment is thus finished, and every question that has arisen in connection with it has been answered, and the whole is fully understood, then write out on the page [or pages] following the preliminary page, a good, clear statement of the experiment, making evident to any future reader the apparatus and material used ; the method followed ; what has been observed ; and the conclusion drawn, or what the experiment has shown. Ink should be used for writing this account if it is desired that the notebook shall have the neatest possible appearance at the end of the year. The preliminary page, however, best be kept in pencil. It has been found a good plan to reserve all left-hand pages for preliminary pages, and all right-hand pages for the account written in ink.

Although some pupils seem to profit by thinking over an experiment for a day or two, after having made their preliminary rough notes, before writing the permanent account, yet there is danger of forming in this way a habit of negligence. In general, the full account should be completed not later than one week from the time the experiment was done in the laboratory.

Represent apparatus, crystalline forms, etc., as far as possible, by sketches. A drawing will often express clearly and forcibly as much as pages of writing. Do not feel discouraged if the first drawings seem failures.

Make good use of the eyes and resolve that each succeeding figure shall be a little better than the one before, and there need be no fear for the appearance of the last.

Be sure to insert an answer in the laboratory-book every time a question is asked in the text-book. Make this answer in the form of a statement in such a way that the answer may be understood by a person who does not know the question. First formulate an answer to every question, and then turn to the instructor and ascertain if the answer is correct, before writing it in the note-book. It seems to the author that every teacher should pay particular attention to this or some similar method of questioning, as it teaches the student to do his own thinking, and gives the instructor a chance to watch the working of the student's mind with a view to giving help to the student in acquiring correct methods of reasoning.

Particular attention should also be given to expressing chemical changes by means of diagrams, as on page 32, but it is *not* advisable to use chemical symbols for abbreviations in these diagrams, even if the student *thinks* he knows the full significance of a chemical symbol.

At Exeter the instructor meets the students, in class, four times a week. In general, at each of the first three of the four periods he assigns some new work to be done in the laboratory, gives any needed directions or precautions, and then holds a review of the work that was assigned one week before. In these reviews the experiments are taken up in their order, the questions in the text are put to the students, and the instructor endeavors to find out, and clear away, any difficulties his students may have met. At the fourth period, which is the last in the week, the instructor has his students write out the fair accounts in ink. While they are writing he passes among them, examines their books, and criticises, — particularly, their manner of recording the work done. This fourth period has come to be, at Exeter, a period of instruction in English composition as well as in chemistry. Between these four periods the students, still under the direct oversight of the instructor, perform their experiments in the laboratory, and make their rough preliminary notes. About 100 hours of laboratory work have been found necessary, by the most careful workers, to complete Part I. Part II requires about fifty hours, and the experimental work of Part III requires about fifty hours.[1] If a student has only a limited amount of time to devote to his chemistry, the author advises him to take Part III after finishing Part I.

[1] It is hoped that students who take up this course in chemistry will have had such laboratory practice in physics that they can omit the actual performance of many experiments of Part III, *e.g.*, "Law of Boyle"; "Weight and Sp. Gv. of Air"; "Law of Dalton"; Experiments 14, 15 and 16 on the Sp. Gv. of gases; " Spaces between the Molecules"; "Irregular Expansion of Liquids"; "Regular Expansion of Gases"; and the experiments under Sp. Heat. In every case of omission, however, reference should be made to the data already prepared and recorded in the note-book of physics, and the results should be carefully reviewed for use in the work in chemistry.

The fifty hours noted above are for a student who has *not* had previous training in Physics.

Part II contains work of the same nature as that of Part I. This additional work, on highly interesting substances, is intended for students who can devote more time to their chemistry than those who are limited to the minimum that is necessary for a general education. It is possible to arrange the work so that the brightest students in a class may do both Part I and Part II while those of average ability are doing Part I, *e.g.*, bright students may be allowed to work ahead of the class, putting only their rough notes in the note-book till the class, in review, has caught up; or two sections of the class may be formed. *But no experiment of Part I should be omitted by any one.*

When the student is in the laboratory he should be careful always: —

To read *through* the description of each part of every experiment before commencing to do that part.

To note carefully all cautions.

To leave balances clean and in proper adjustment.

To keep all weights clean.

To resist the temptation to use weights for any other purpose than weighing.

To avoid putting test papers in liquids. [See Appendix C.]

To use only clean apparatus.

To put on the preliminary page of the note-book only *brief* notes, and notes right to the point.

To use the utmost care in all work with hydrogen gas.

And at all times : —

To use his utmost good judgment, and if in difficulty or in danger, to try to *think* his way out clearly and quickly.

Chemicals and apparatus should be provided well in advance of the time they will be needed, in order that there may be no loss of time in waiting for suitable apparatus or the necessary chemicals.

Mr. M. A. Buck, at 5 Tremont Street, Boston, is prepared to furnish apparatus and chemicals suited to the needs of students who pursue this course. On application, he will furnish carefully prepared lists of all apparatus and chemicals needed, either for the whole book or for any part, — for a liberal allowance, or for the minimum amount required by careful students, — and his arrangements are such that he can usually furnish the articles, whether ordered in complete sets or individually, at prices below the dealers' regular quotations.

As Mr. Buck was an assistant in the Exeter Laboratory for several years during the development of this course, the author feels sure that he will prove well fitted to furnish *exactly the right articles* to make the work run with the least amount of friction from " misfit " apparatus and chemicals unsuited for the work in hand.

PRELIMINARY WORK.

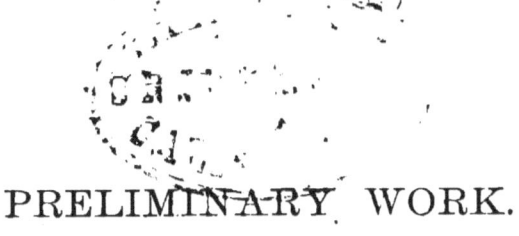

PRELIMINARY WORK.

IN THE LABORATORY.

I. *MEASURING.*

HAVE ready a metric rule, about thirty centimeters long and graduated to millimeters; a rule about one foot long and graduated in inches and fractions of inches, down at least to one eighth inch;[1] a glass cylinder, capacity 50 or 100 cubic centimeters, and graduated to cubic centimeters; a graduate holding one fluid ounce; several glass beakers, flasks, and bottles of different sizes; a test-tube rack containing five or six test-tubes of different sizes.

Measure, first in the metric system, then in the English, the length of your laboratory desk; also its width; estimate its area in square centimeters, in square inches, in square millimeters, in square meters. In recording measurements use the decimal point, *e.g.*, if the length of your desk is one meter, and two decimeters, and five centimeters, and seven millimeters, do not record the measurement 1^m, 2^{dm}, 5^{cm}, and 7^{mm}; but express it 1.257^m, or 125.7^{cm}, or the like. How many centimeters are there in one inch? How many inches in one meter? How many square centimeters in one

---

[1] A convenient form of rule is one having centimeters on one side and inches on the other.

square inch? What fraction of a square inch does a square centimeter occupy? Put the answers to these questions in your note-book. Also fix the round numbers in your mind.

Take your graduated cylinder [commonly called "graduate"], fill your largest flask to the brim with water, pour the water from the flask into the cylinder, and find how many cubic centimeters the flask holds. Find also the capacity in cc [cc stands for cubic centimeters] of your other flasks, beakers, bottles, and test-tubes.

One purpose of this work in measuring is to enable you to tell at a glance about how much the laboratory vessels in common use contain. It is best, then, in making these measurements, not to fill the beakers to the brim, but conveniently full only. Try to fix the various capacities in mind.

Also find the capacity of at least two bottles, your smallest beaker, and a test-tube in fluid ounces, using your oz. graduate as a measure. Fix the values in mind as well as record them in your note-book. Bottles are commonly designated, in trade, by the number of ozs. they contain.

II. *WEIGHING.*

Have ready a platform balance, capable of holding a load of 5 kilos and sensitive at least to one tenth of a gram; a set of iron weights, $2000-10^g$ inclusive [g stands for gram, sometimes written gramme]; a smaller balance, sensitive at least to one one hundredth of a gram; a set of weights, 50^g-10^{mg} inclusive [mg stands for milligrams], accurate at least to one one hundredth of a gram; a pair of brass forceps for handling the weights.

Take the large balances, fill three of your larger vessels conveniently full of water; weigh each separately, then all three together. Weigh accurately to a single gram. See if the sum of the three separate weights equals the combined weight. Record *all* observations in your note-book. In recording use the decimal point as in the case of your measurements.

Balance your graduate on one pan with any convenient article, as, for instance, a vessel partly filled with water or, better, with lead shot.[1] Pour water into the graduate to the amount of exactly 50^{cc} [100^{cc} if you have a 100^{cc} graduate]. Get the weight of this water and notice that one cc of water weighs exactly one gram.

Take the smaller balances, and, using the smaller weights [which must be handled with the forceps only], weigh [accurately to centigrams[2]] three nails, first separately, then all together. The sum of the separate weights should not vary from the combined weight more than 0.05^g. If the variation is greater, repeat all the weighings. Record results showing the variation, if any.[3]

---

[1] It is a good plan always to have at hand in the laboratory two or three saucers or similar stout vessels filled with shot, to be used for *balancing* vessels whose contents only are to be weighed accurately.

[2] If you are not familiar with the denominations of the metric system, look up this system in some arithmetic or in the dictionary.

[3] Unless perfectly familiar with the English system of weights and the relation between these weights and the metric, obtain a set of English weights, and with these weigh all the articles you have

III. *MAKING A WASH-BOTTLE.*

Take a 500^{cc} flask and a cork to fit; also about 1 meter of soft glass tube with a bore of about 6^{mm} [mm stands for millimeters]. Put the cork on the floor and with the sole of the shoe roll it to soften it and make it fit the neck of the flask tightly. With a rat-tail file bore two holes through the cork. Let each hole be *round* and of such size that the glass tube will fit it tightly. Take a piece of the glass tube about 10^{cm} [cm stands for centimeters] longer than the height of the flask.[1] Stand the glass tube in the flask, make a mark on the tube half way between its upper end and the top of the flask.[2] Bend the glass tube[2] at the mark until the shorter limb forms with the longer an angle of about 45°. The bend should form a curve, not a sharp angle. Pass the longer limb through a hole of the stopper, then at a point about half way between the first bend and the open end of the longer limb, make another but lesser bend. This latter bend should bring the lower end of the tube, when the cork is inserted in the flask, to the vertex of the angle made by the bottom of the flask with its side. Be sure to make the second bend *toward* the first, i.e., in making the second let the open end of the longer limb *approach* the open end of the shorter limb — not recede still

---

weighed with your metric weights. Compare results, and note that one ounce is equal to about 28.3 grams and that one pound contains about 454 grams.

[1] To cut glass tubes, see Manipulations [at end of book].
[2] To mark tubes, see Manipulations.
[3] See Manipulations.

farther from it. Now take a piece of glass tube about 15 centimeters long, bend this slightly, best till the two limbs form an obtuse angle of 135°. Take a third piece of glass tube and draw it out[1] to a diameter of about 2^{mm}. From this drawn-out tube cut a tip for the wash-bottle. At one end this tip should be of the same diameter as the original tube; at the other, not over 2^{mm}; in length, $3-5^{cm}$. Wipe all soot from the three tubes. Fire polish[2] every rough end of the tubes. Insert the second bent tube through its hole in the cork, letting the end be flush with the lower end of the cork. This tube forms the mouth-piece. Attach, by means of a rubber connector, *i.e.*, a piece of small rubber tube $1-2^{cm}$ long, the tip to the shorter end of the longer bent tube. This tip forms the jet, and, owing to its flexible connection of rubber, can be directed by the fingers in different directions when the wash-bottle is used for washing precipitates. The slight inner bend in the longer tube enables the whole of the water to be blown from the bottle when the flask is tipped up, as it usually is, in use. Fill the bottle about two thirds with water. [Distilled water[3] is best, though not necessary, for the following experiments. When necessary, the fact will be stated.] Wet the cork to fill the pores, insert the cork in the flask, and blow through the mouth-piece. The tip should deliver a fine, steady stream. Insert a sketch of your wash-bottle in your note-book.

[1] See Manipulations.   [2] See Manipulations.   [3] See Appendix O.

PART I.

EXPERIMENTS.

PART I.—EXPERIMENTS.

Experiment 1.

Iron.—A *solid* substance.

A. The Properties of Iron.

TAKE an iron nail and a piece of fine iron wire. Note as many of the properties of iron as you can, *e.g.*, color [make a fresh scratch with a file to get the true color]; hardness [see what it will scratch, and what will scratch it,—try, for instance, glass, chalk, wood, a file, *etc.*]; tenacity [try pulling apart the nail, the wire]; brittleness [mention some things you find, on trying, to be more brittle, some less]; fusibility [try to melt the wire, first in the flame of the Bunsen burner,[1] then in that of the blast-lamp]; volatility [see if any heat you can produce will make it go off in vapor in the way steam does from hot water].

B. Action of Air on Hot Iron.

Fill a small porcelain crucible about half full of iron filings. Weigh carefully [to centigrams]. Set the crucible on a pipe-stem triangle supported on a stand. Heat with a Bunsen burner for about ten minutes with occasional stirring [with a glass rod], that the air may come in contact with all the filings. Cool the crucible.

[1] For the use of burners, see Manipulations.

Why? Again weigh. What has caused the gain? See if you can detect any difference of lustre between filings not heated in air and those heated. Sprinkle a few filings in the flame and note the phenomenon. What is a phenomenon? Do not return filings once used to the bottle. Let us call that which has come from the air and fastened itself to the filings — **oxygen,** and the new dull-black substance formed — **oxide of iron.** Let us call oxide of iron a **compound,** because it is obviously compounded of two other substances. Let us call iron itself a **simple substance,** because we cannot make it from two or more other substances, nor can we get two or more other substances from it. What is a simple substance? What is a compound?

Having found that there is in the air a peculiar substance capable of joining iron, it becomes of interest to ascertain what proportion of the air this substance occupies. In our investigation we shall need the aid of another simple substance, phosphorus, with which the oxygen unites even more readily than with iron.

Experiment 2.

Phosphorus.—Another *solid* substance.

A. **The Properties of Phosphorus.**

Take a piece of yellow phosphorus and a little red phosphorus.

Caution! Caution! Caution!

Yellow phosphorus is very *poisonous* indeed, extremely *inflammable*, and a phosphorous *burn* is very painful. This substance must be stored *under water*, and cut only under water, — best in the pneumatic trough or in a large basin. Matches must not be kept in closets or drawers.

Examine the phosphorus and note its most important properties, as color, consistency, fusibility, inflammability, *etc.* Do not handle the yellow with bare fingers. Use iron forceps. Dry it rapidly by pressing it gently between filter or blotting papers. In stating the properties, make two tables: one for the red, the other for the yellow.

B. Action of Air on Warm Phosphorus.

Have ready a dry,[1] *quick-sealing* pint fruit-jar,[2] and a clean iron deflagrating spoon, also about half a gram of phosphorus. If you use the yellow phosphorus it must be cut under water, dried quickly by pressing between filter papers, and at once placed in the spoon. Have at hand a Bunsen burner flame, over which the spoon may be held in order to light the phosphorus.

The rubber washer to the jar should be greased [vaseline is good] to make it tight. Hold the spoon in one hand and the cover to the jar in the other. Light the phosphorus, and with a quick but deliberate motion plunge the spoon in the jar, at once put on the cover, fasten it and step back, as the jar may crack.

Note, as the phosphorus burns away, the white powder formed. As soon as the jar cools, open it,

---

[1] To dry flasks, jars, and other pieces of apparatus, see Manipulations.

[2] A quick-sealing fruit-jar with a rubber washer. Those called "Lightning" are excellent.

holding the mouth under water, which, rushing in, will show that a part of the air has gone. Make a rough estimation of what part [by volume] has gone. This is the same part that, in Ex. 1, B, left the air, joined the iron, and increased the weight. Remember that we are going to call this gas which has the power of joining other things, oxygen. We will call the new substance made from the oxygen and the phosphorus [the white powder that formed in the jar] **oxide of phosphorus.**

Note that not all the air was used,—part was left. This part is another gaseous substance which we shall study later.

Experiment 3.

Mercury.—A *liquid* substance.

Caution! Be careful in the use of heat with mercury, as the vapor of mercury is a vigorous poison. Take a globule of mercury as large as a pea and note its chief properties. In doing this, review Ex. 1, A, and Ex. 2, A. Note those respects in which mercury resembles iron or phosphorus, and those in which it differs from these substances.

Experiment 4.

Carbon.

Take a bit of charcoal, a bit of graphite [from a "lead" pencil], some soot, a bit of gas-retort carbon, and [if obtainable] a diamond. Also burn a piece of thin paper to get the carbonaceous residue, which often keeps the original form. Study these different forms of carbon and note chief properties. What is meant by allotropic forms?

Experiment 5.

Artificial Preparation of Oxygen.

A. From the Red Oxide of Mercury.

Take a piece of hard glass tube, and red oxide of mercury. **Caution!** Red oxide of mercury is a vigorous poison. *Note.* Red oxide of mercury may be made by prolonged heating of mercury in contact with air. Recall the preparation of the black oxide of iron that you made in Ex. 1, B. If the red oxide of mercury itself is heated vigorously, it is separated into the mercury and the oxygen from which it was made.

Make a small matrass[1] from hard glass tube. Fill the bulb about one half with the red oxide of mercury, and heat over the Bunsen burner flame. Test for oxygen by plunging down the tube a bit of glowing [not flaming] carbon [best made from a splinter of wood or a burned match]. What did we call the substance formed in Ex. 1, B, when oxygen joined the iron filings? what the substance formed from the union of oxygen and phosphorus when, in Ex. 2, B, the phosphorus burned? What, then, shall we call the substance formed when the oxygen now joins the carbon, causing the intense heat and fire at the end of the splinter? What becomes of this new substance which

[1] See Matrass, under Manipulations.

is formed? Note the globules that collect in the cool part of the tube. Break open the tube and examine them. What are they?

This process, by which a compound substance is split into simpler ones, is called **Analysis,** and is typical of a vast number of chemical changes.

B. From Chlorate of Potassium.

Note. As red oxide of mercury is expensive, some other method than that of part A is desirable for preparing oxygen on the large scale. It is found that chlorate of potassium, when heated to a high temperature, is also broken up into two substances, one of which is oxygen.

Have ready a Kjeldahl flask[1] clamped to a stand at such a height that the body of the flask may be heated conveniently by a Bunsen burner. Fit the flask with a one-hole cork, and from the cork let a glass tube, of not less than 5^{mm} bore, pass down into a trough of water.[2] The end of the delivery tube[3] should be turned up a little in the water. Soak the cork for a minute or two to fill small holes.

Put in the flask about 15^g of chlorate of potassium. Weigh the flask with its charge. Do not have the cork in when you weigh. Clamp the flask in position. **Caution!** In this experiment never insert the cork tightly, as the tube may plug up, and if the cork cannot

---

[1] A stout, long-necked, pear-shaped flask of hard glass.

[2] See Pneumatic Trough, under Manipulations.

[3] The term "delivery tube" will appear frequently hereafter, and is to be taken to mean a tube for the delivery of a gas from a flask, bottle, or the like.

blow out [like a safety valve], the flask may explode. Have ready four pint jars filled with water and standing inverted on the bridge of the pneumatic trough. What does pneumatic mean? Heat the chlorate of potassium with a Bunsen burner. **Caution!** Do not allow the hand to come directly under the flask, for if the flask should crack and the hot, melted chlorate run out, the hand might be scalded. Move the flame all around on the bottom of the flask in order that no part of the glass may get too hot and soften. Catch the gas evolved. Catch at least four jars full. Not more than a few cc of water should be left in a jar. Snap on the covers while the jars are still inverted with their mouths under water. Use rubber washers, well greased, when sealing the jars. Set the jars of gas away for further use. *Uncork your flask* before you stop heating it. Why? Weigh the flask with the residue. Compare weight with first weight. Explain the change in weight. Take your first-caught jar of gas, and prove that the gas is similar to that from the red oxide of mercury, *i.e.*, that it is oxygen. Prove this by plunging in a bit of glowing carbon.

Experiment 6.

A. **Action of Undiluted Oxygen Gas on Iron.**

Have ready a jar of oxygen gas [from Ex. 5, B], also a small amount of very fine iron filings. Two minutes' continuous filing of a board nail with a five or six-inch file over glazed paper furnishes an ample amount of good filings. Those used in trade often are not fit for this experiment. Place the filings in a little heap in a *clean* and *dry* deflagrating spoon.[1] Heat them [spoon and all] over a Bunsen burner for a moment till a dull glow runs through the heap; then plunge them at once into the jar of oxygen and put on the cover. Compare the action of pure oxygen gas with that of the oxygen in the air [see Ex. 1, B], which is diluted with nearly eighty per cent [by volume] of another gas. Save the oxide of iron for Ex. 15.

B. Action of Undiluted Oxygen Gas on Phosphorus.

Proceed as in Ex. 2, B, except that the phosphorus is to be burned in a jar of pure oxygen and greater precautions in every way are to be taken, as the action is violent and the danger of the jar exploding much greater. Weigh the phosphorus exactly. In no case use more than $.55^g$ for a pint jar. It is best to throw a cloth around the jar after the action and keep the cloth around till the jar is opened under water. This will catch pieces of glass if there is an explosion. What becomes of the white oxide of phosphorus in this experiment?

C. Action of Undiluted Oxygen Gas on Carbon.

Proceed as in B, but use a lump of charcoal weighing about 1^g instead of the phosphorus. Get the charcoal well on fire by heating it with a Bunsen burner

---

[1] If the bottom of the deflagrating spoon is too thick, the filings will not become heated enough by the Bunsen burner before they are coated with oxide, and in this case the experiment will be a failure.

and *seal the jar as soon as possible.* Why? Do not use a deflagrating spoon that has any phosphorus in it. Test the spoon by holding it a minute in the Bunsen flame.

A pretty effect may be obtained if a little *powdered* charcoal is sprinkled over the lump of charcoal.

If on opening the jar under water the oxygen does not appear to have been used, test the remaining gas with a glowing match. Give the chief properties of oxide of carbon. This addition of oxygen to another substance is called **oxidation**. The term is also sometimes applied to the addition of other substances.

Experiment 7.

Preparation of Oxygen from Water.

Have ready some water in a suitable vessel[1] into which pass two platinum electrodes so arranged that the electric current in passing from one electrode to the other must pass through the water. Pass a current of electricity through the water from one electrode to the other. If the resistance of the water to the passage of the electric current is too great for a rapid evolution of gas, add a little sulphuric acid. In some way the acid greatly helps the current to get through the water. Fill a small tt with water. Place your thumb over the mouth of the tt, and invert the tube over the electrode that is giving off the gas the *slower.* Catch a tt of the gas. Do not let any of the other gas get in. Test the

[1] See Appendix A.

contents of the tt [glowing match test], and compare with the gas got from the air [see Ex. 1, B and Ex. 2, B]; also with the gas from red oxide of mercury [see Ex. 5, A]. Give all the properties you can of oxygen.

Experiment 8.

Examination of that Constituent of Water which is not Oxygen.

Again pass the electric current through water, as in Ex. 7. This time collect the gas which is given off in larger quantity. Catch a tt full. Test this with a glowing match, still holding the tt upside down. Why upside down? Answer this after doing Ex. 9, B. Is the gas oxygen? If not found to be oxygen, test at once with a flaming match. Again catch some of the gas, — about two-sevenths of a tt this time. Carefully let the air fill the rest of the tube. Put your thumb over the end, and shake to mix the air with the gas. Now apply quickly a lighted match. What takes place chemically? Let us call this new gas **hydrogen**.

Experiment 9.

A. The Preparation of Hydrogen from Water by Means of Iron.

Note. Ex. 1, B, taught us that iron when hot has attraction for oxygen. If water [best in the form of steam] is passed over red-hot iron the iron will decompose the water, take the oxygen to itself and leave hydrogen. What must then happen to the iron? Proceed as follows, and see if your inference as to the answer to this question is confirmed by experiment. Take a piece of half-inch gas pipe about two feet long with a one-hole cork in each end. Put about 30^g of iron filings[1] in the gas pipe, as near the middle as possible, but be sure there is a passage through the whole tube. Clamp the gas pipe to a stand at each end. Heat the filings by means of two Bunsen burners directed at the same point on the gas pipe. Fit your wash-bottle flask with a one-hole cork and delivery tube. Set the flask on a tripod stand on which is a piece of fine iron gauze about four inches square, to prevent the bottom of the flask being unevenly heated. Put some water in the flask, set a Bunsen burner below, generate steam and pass the steam over the hot filings. Have a delivery tube passing from the filings down into the pneumatic trough. Let all corks be tight. Catch several tts of the gas evolved. Do not try to keep it corked up, but catch the gas when you need it for the following work. Reject the first two tubefuls because there is apt to be air in them.

Caution! Caution! Caution!

Hydrogen and air form an extremely dangerous explosive mixture. Use the utmost care in all work with hydrogen. Think what you are going to do in every case before you act.

Test the third and fourth tubes, as you tested in Ex. 8, to prove that this is the same gas we agreed to call

---

[1] The filings should be as free as possible from dirt and oil, otherwise a troublesome smoke may appear.

hydrogen. Note all phenomena as you test. How do you explain what happens when you apply the flame?

What shall we call the substance that results when the oxygen of the air joins the hydrogen, causing the intense heat and fire? What is the common name for this substance? As in making oxygen from chlorate of potassium, here uncork your wash-bottle flask before you stop heating. Why?

B. Specific Gravity of Hydrogen.

Catch a tt of hydrogen. Put your thumb over the end of the tube. Remove the tube from the trough. Hold the tube *upside down*. Remove the thumb carefully. Wait while you breathe naturally ten times, *i.e.*, wait about half a minute. At once apply a flame to the tt's mouth. Again catch a tt full and proceed as before, but now hold the tt, with its mouth up, while you breathe ten times. Again catch a tt full. Hold a second tt bottom side up and carefully pour the hydrogen from the other *up* into the second. At once test the contents of the second for hydrogen. What do you say in regard to the weight of hydrogen?

C. Action of Air on Warm Hydrogen.

Have ready an eight-ounce, wide-mouth bottle fitted with a two-hole stopper. Let there be projecting up straight from one hole a piece of hard glass tube about six inches long with one end just reaching through the cork, and with the other [upper] end drawn down to an opening about like the tip to a wash-bottle nozzle. Through the second hole of the cork a piece of common glass tube should be inserted reaching well into the bottle and connecting the bottle with the hydrogen gas pipe. This bottle serves as a catch-bottle or trap to condense any steam that gets by the filings. It best be set in a dish of cold water or, better, in the pneumatic trough. Generate hydrogen as in A, but apply the blast lamp for a few minutes that the filings may be heated red-hot and cause a good flow of hydrogen. As there is some danger of the catch-bottle blowing up, apply to the instructor [1] for a method of testing the explosive quality of the contents of this bottle. When all is safe, light the hydrogen as it issues from the small jet, and let it burn in the air. What is the burning? What the product of combustion? Hold a dry and cold tt over the jet. Do not smother the flame, however. Note the substance that forms on the sides of the tt. What is it?

Experiment 10.

Sulphur.

A. The Properties of Sulphur.

Take some roll brimstone, and some flowers of sulphur. Note the chief properties of sulphur. Review the records of iron, phosphorus, mercury, carbon, oxygen, hydrogen, and all the oxides you have made. Compare sulphur with the other substances you have studied.

[1] See Appendix B.

B. Modifications of Sulphur.

1. Dissolve about a gram of roll brimstone in about five cc of sulphide of carbon [one of the very few things in which sulphur will dissolve]. **Caution!** *Sulphide of carbon is very volatile and* **inflammable.** *Have no fire near.* Best grind the sulphur to a powder in a mortar to make it dissolve quickly. The flowers do not dissolve as well as the roll. Pour the solution out in a crystal pan,[1] and let the sulphide of carbon evaporate spontaneously. Examine the form and color of the crystals deposited.

2. Melt enough roll brimstone to fill a small common beaker nearly full. In melting the sulphur, the beaker should be set on iron gauze or asbestos paper. If the sulphur in the beaker catches fire turn off your gas and smother the fire with an inverted dish or a cloth. Take care that the temperature does not rise much above the melting point of the sulphur. Let the sulphur cool till crystals begin to shoot across the surface and just meet in the middle, then promptly pour out into water what remains liquid. Note the form and color of the crystals left *in the beaker*. Compare with those of 1.

3. Melt some sulphur in a tt. Hold the tt over a naked Bunsen flame.[2] Raise the temperature till the substance, after it melts to a liquid, becomes thick and viscous. What is meant by viscous? Then pour the sulphur out in a fine stream into cold water. Note the form of the sulphur in the water. Compare sulphur

---

[1] A *shallow* glass dish.
[2] See Manipulations for method of holding a hot tt.

with carbon, phosphorus, *etc.*, in regard to allotropy. What is allotropy?

C. Action of Oxygen on Hot Sulphur.

Prepare more oxygen in the following manner. Take 12^g chlorate of potassium and mix with it 3^g of powdered black oxide of manganese.[1] The black oxide should be mixed intimately with the chlorate. Do not spill any. Heat the mixture in a Kjeldahl flask as in Ex. 5, B. Do not use here a pneumatic trough filled with water, for collecting the gas, but fill three or four *dry* jars *by displacement, i.e.*, pass the delivery tube [best have one ending in about 15^{cm} rubber tube,] directly into the jar. Hold the cover on as well as possible. The oxygen gas, which is somewhat heavier than air, will collect at the bottom of the jar and push the air up and out. You can tell when the jar is full by holding a glowing match at the crack left between the cover and the edge. When the jar is *full*, withdraw the rubber tube slowly that the oxygen may fill the space occupied by the tube. Snap on the cover at once. Set away, for the next experiments, at least two jars that seem perfectly dry. Do not throw away the contents of the Kjeldahl flask, but, when cool, add about 100^{cc} of warm water. The white chloride of potassium dissolves, while the black oxide of manganese does not. Take a funnel and a filter paper[2] to fit the funnel. Pour the contents of the flask on the filter

---

[1] The black oxide of manganese of trade is sometimes adulterated with coal dust. Such adulteration might cause a serious explosion in this experiment. Why?

[2] For directions about filter papers, see Manipulations.

paper. By means of the wash-bottle pass about 100^{cc} of water through the black oxide of manganese to carry through the chloride of potassium. Dry the black oxide and weigh it. In order to dry the black oxide, have ready a weighed porcelain evaporating dish. Transfer the black oxide to this dish, using a fine stream from the wash-bottle to wash the last traces of the oxide into the dish while you hold the paper just above the dish. Evaporate off the water, *avoiding all spattering*,[1] and with a small flame dry the black oxide in the dish **to constant weight**. In drying, never allow the black oxide to be heated to redness. Why? To dry a substance to constant weight, heat it until it seems dry, then weigh it, again heat it, weigh again, and if the weight is the same as that previously found no further drying is necessary. If there has been a loss after the second heating, again heat, weigh and so continue till no further loss in weight is found. You should have the same amount of black oxide that you started with, *i.e.*, 3^g.

Note. This black substance is the oxide of the metal manganese. Compare it with the oxide of iron you made. We do not know its action in this experiment, but it certainly makes the oxygen come off easily. What do *you* think of its action? Compare the use of sulphuric acid when you decomposed water with the electric current.

---

[1] Spattering is best avoided by evaporating at such a low temperature that the liquid does not actually boil. Best set the evaporating dish over a beaker containing water which is kept boiling by a Bunsen burner beneath. This method of slow evaporation is called "Evaporation over the water-bath."

Burn a small piece of sulphur on a clean deflagrating-spoon in a jar of oxygen. Open under water, note the condition of the jar, and at once snap on the cover again. If no water has entered the jar, let in about 50^{cc}. Shake with the water that has entered. Again open under water and note the condition of the jar. Note the properties of the new compound formed, especially its state, color, odor, and solubility in water. What shall we call the new compound? Will it weigh more than the original oxygen?

Experiment 11.

Sulphurous Acid.

Again burn sulphur in a jar of oxygen. Do not open under water, but remove the spoon carefully, and add about 20^{cc} of water. Shake well. Try the effect of the liquid on a bit of blue litmus paper.[1] Try the effect of water itself on the same kind of test paper. Taste a very small amount of the new liquid. Let us call our new substance **sulphurous acid**. What three simple substances must there be in this acid?

Experiment 12.

A Second Oxide of Sulphur.

Oxygen may be made to join the gaseous oxide of sulphur and form a second oxide of sulphur. Have

[1] For the preparation and use of litmus paper see Appendix C.

ready weighed in a clean and dry deflagrating-spoon just 0.3^g of sulphur. Burn the sulphur as in Ex. 11, but do not remove the spoon. The 0.3^g of sulphur are not enough to use all the oxygen. Therefore you have in the jar, after the burning, oxygen and oxide of sulphur. What are the properties of oxygen, — of oxide of sulphur? Have ready a suction pump[1] and a tube of hard glass containing a little platinum sponge, or platinized asbestos.[2] This tube containing the platinum should be 15–20^{cm} long and have a bore of about 6^{mm}. The platinum sponge, or asbestos, should not be packed so tightly that the gases cannot easily pass through.[3] Also have ready a piece of glass tube bent in the form of a U, the total length of this tube to be about 40^{cm}. One end of this tube should be connected with the platinum sponge tube, and the other with the suction pump — each by a piece of rubber as short as possible, for the oxide formed corrodes rubber. Place the U-tube in a freezing mixture.[4] Suck dry air through the whole apparatus, slowly, for five minutes, to remove all moisture. The air may be dried by connecting the platinum sponge tube with two catch-bottles[5] of sulphuric acid. Heat the platinum sponge well with a Bunsen burner. The sponge tube may best be supported by a stand and wire gauze. As soon as the

---

[1] For the use of the suction pump see Appendix D.

[2] 0.1^g of platinized asbestos is sufficient, but 0.2 are better.

[3] It is well to insert within the hard glass tube two bits of small soft glass tube — one on each side of the platinum — in order to prevent the platinum being driven out by any sudden puff of gases.

[4] Crushed ice or snow, salt, and enough water to make a pasty mass, are good for a freezing mixture.

[5] See Appendix E.

sponge is hot, and the whole apparatus dry, disconnect the catch-bottles, fit a piece of rubber tube about 15^{cm} long to that end of the platinum sponge tube from which the catch-bottles were removed, pass the end of this rubber tube to the bottom of the jar containing the mixture of oxygen and the gaseous oxide of sulphur, and then, using the hand to prevent, as far as possible, the air mixing with the contents of the jar, let the pump *slowly* suck the gases from the fruit jar, over the hot platinum, into the cooled U-tube. Look for white crystals in the U-tube, which must be kept very cold.

If you should weigh the platinum sponge before the experiment and after, you would find no change in weight. The sponge itself does not make any part of the crystals. From what must the crystals be made? Remove the crystals from the cold bath and examine them before they melt. Let us call this new substance the **second oxide of sulphur.** Keep the new substance for the next experiment.

Experiment 13.

Sulphuric Acid.

Compare the record of Ex. 11, where you made sulph*ous* acid. Now take the second oxide of sulphur made in Ex. 12 [which becomes liquid if allowed to stand long at ordinary temperature], and add a few drops of water from the wash-bottle. Try the effect of the resulting liquid on blue test paper. Dilute more and taste of a very little. Let us call our new acid **sulphuric acid.** What three things, each simple, must there be in sulphuric acid? In what respect does sulphuric acid differ from sulphurous?

Experiment 14.

Removal of Hydrogen from Sulphuric Acid.

Take a small flask fitted with a one-hole cork and a delivery tube reaching to a pneumatic trough. Pour about 10^{cc} of water into the flask. Then add 5^{cc} of sulphuric acid.[1] **Caution! Never add water to sulphuric acid.** *Add the sulphuric acid to the water.* Sulphuric acid and water can generate great heat. If the amount of acid is large and that of water small, this heat may boil the water with explosive violence.

Have ready about 10^g of iron [best in the form of small nails]. Add the iron to the flask before the liquid has had time to cool, and insert the cork with its delivery tube. **Caution!** Keep all fire away. Why? After the gas has passed long enough to drive the air from the flask, catch a tt of it and test it. What is it? Prove that iron has not the power of removing the hydrogen from water, at a low temperature, by putting some nails in a tt and warming till the water is at

---

[1] When the student has once made a compound substance, it is not supposed that he is to make enough for all subsequent work. He should be supplied with the commercial article.

least as warm as was the mixture in the flask. If the hydrogen did not come from the 10^{cc} of water, whence must it have come? What has happened to that part of the sulphuric acid that was made of sulphur and oxygen? Answer this question by experiment, as follows:

Put the contents of the flask in an evaporating dish. Add about 25^{cc} of water. Set the dish on tripod or ring, and warm *gently* till no more hydrogen is given off. [While warming, keep the volume of the liquid as nearly constant as possible by adding water if any evaporates off.] Filter and evaporate the liquid till a little of it, when taken out in a tt and cooled, will deposit crystals. Then at once, while still hot, again filter the liquid into a beaker. Hold the beaker so that cold water shall flow over the outside till the liquid within is cold. Note the crystals formed. Filter, spread the crystals on paper to dry. Note their properties. Let us call the new substance **sulphate of iron.** The green crystals are made of sulphate of iron and **water of crystallization,** *i.e.*, water which is in some way joined to the sulphate of iron. Many substances in this way form crystals by the addition of water. Put some of the green crystals in a dry tt and heat over a Bunsen burner. Note the formation on the sides of the tt. What forms there? Of what simple substances do you say sulphate of iron is composed? Of what compound substances were the green crystals composed?

Iron is a simple substance, and sulphuric acid, we have proved, contains hydrogen, sulphur, and oxygen.

The mutual action of iron and sulphuric acid may be represented by a diagram, thus: $\begin{array}{|c|c|}\hline \text{iron} & \text{hydrogen} \\\hline \text{sulphur} \\ \text{oxygen}\end{array}$. This indicates that the iron has changed place with the hydrogen of the acid, *i.e.*, the hydrogen has become free, and the iron has become combined with the sulphur and oxygen that were in the acid, making a new substance —iron sulphate—whose composition may be represented thus: $\begin{Bmatrix}\text{iron}\\\text{sulphur}\\\text{oxygen}\end{Bmatrix}$.

Remembering that water is oxide of hydrogen, *i.e.*, hydrogen + oxygen, we may represent the formation of the green crystals thus:

$$\begin{Bmatrix}\text{hydrogen}\\\text{oxygen}\end{Bmatrix} + \begin{Bmatrix}\text{iron}\\\text{sulphur}\\\text{oxygen}\end{Bmatrix} \text{ gives (or } =) \begin{Bmatrix}\text{iron}\\\text{sulphur}\\\text{oxygen}\\\text{hydrogen}\\\text{oxygen}\end{Bmatrix}$$

Note. We have already proved that sulphuric acid contains more than one portion of oxygen: hence "oxygen" in our $\begin{Bmatrix}\text{hydrogen}\\\text{sulphur}\\\text{oxygen}\end{Bmatrix}$ stands for the total amount of oxygen present in the acid. How have we proved that there is more than one portion of oxygen in sulphuric acid?

Note. Having found that water added to either oxide of sulphur forms an acid, it becomes of interest to see if water can form an acid with any other oxide except the oxides of sulphur.

Experiment 15.
Action of Water on Oxide of Iron.

Add a little water to the oxide of iron made in Ex. 6, A. Test with litmus paper. Has an acid been formed?

Experiment 16.

Action of Water on Oxide of Phosphorus.

Again prepare some oxide of phosphorus, by burning a small amount of phosphorus in a *dry* jar of air, or, better, oxygen. To the oxide add about 10^{cc} of water. Shake. Test the properties of the resulting liquid. Taste of a drop. How many and what *simple* substances have been used in the preparation of **phosphoric acid**?[1]

Experiment 17.

Action of Water on Oxide of Carbon.

Again prepare some oxide of carbon. *Be sure, by testing, that all vessels used in this experiment are free from sulphuric acid.* To the jar containing the oxide of carbon add about 10^{cc} of water. Shake well. Test the resulting liquid both by litmus and by taste. Compare its acid properties with those of the other acids you have made. Which acid has the most marked acidity? Which the least? Let us call the acid substance made in this experiment **carbonic acid**. What simple substances make up carbonic acid? Place the carbonic acid in a small beaker and warm. What happens? Finally bring the liquid just to a boil. Test the liquid remaining in the beaker with litmus.

[1] Phosphorus is capable of forming several acids: the one made here is commonly called metaphosphoric acid.

What is the liquid left? Are the component parts of carbonic acid bound together strongly or not? What are the component parts of carbonic acid?

Experiment 18.
Zinc.

A. The Properties of Zinc.

Take some zinc,—sheet, granular,[1] and dust. Get the chief properties of this substance.

B. Oxidation of Zinc.

Put a few grams of zinc in a small Hessian crucible. Heat over the blast-lamp till the zinc *melts*. Continue the heating, with an occasional stir by means of an iron rod, till the zinc *burns*. What is the burning chemically? Examine the product of the combustion, which sometimes forms what is called Philosophers' Wool. Get the properties of oxide of zinc, especially its color when hot—when cold. Save some zinc oxide [free from zinc] for C. Having found that many oxides with water form acids, it becomes interesting to add water to every new oxide we make.

C. Action of Water on Oxide of Zinc.

Take a little of the zinc oxide made in B, put it in a tt and add water. Note effect. Test the liquid with litmus,

---

[1] Granular zinc [the most convenient form for general chemical use] may be made by melting any form of zinc in a ladle, Hessian crucible, or other suitable vessel, and pouring the molten metal from a height into a vessel of water.

particularly with litmus that has been turned *slightly* red by a weak acid, as carbonic. Compare [and state the result of comparison] the action of water on oxide of zinc with action of water on other oxides you have tried.

Experiment 19.

Action of Zinc on the Non-combustible Oxide of Carbon; or,

Preparation of a Second Oxide of Carbon.

Have ready a piece of large, hard glass tube about 20^{cm} long. Put in the midst of this tube four or five grams of zinc in the form of zinc dust.[1] Also have ready a rubber bag nearly filled with the non-combustible oxide of carbon.[2] Clamp the glass tube with its charge of zinc at a convenient height for heating the zinc with a Bunsen burner. Fit an empty gas bag to one end of the tube and the bag containing the oxide of carbon to the other. Heat the zinc and slowly pass the gas from bag to bag for a few minutes. Note the formation of oxide of zinc — yellow when hot, white when cold. Whence came the oxygen to form oxide of zinc? Does the resulting gas occupy as much space as the original gas? Remove the bag from the tube without letting any of the gas escape. Fit a cork, through which passes a short piece of glass tube, ending in about 15^{cm}

---

[1] The zinc dust should be in the form of a *very* fine *powder*.

[2] This gas should be prepared by the instructor and given to the student whenever needed up to the time of doing the experiment in which the student learns the action of an acid on marble. For a method for preparing this gas on a large scale, see **Appendix F**.

rubber tube, to the mouth of the rubber bag. Catch a small bottle, *e.g.*, a two-ounce salt-mouth bottle, full of the gas over the pneumatic trough. Do not use more than one half of all, as some must be saved for Ex. 20. Set fire to the new gas in the bottle. Have a dark background, as the new gas will burn with only a pale flame. Note the color of the flame. **Caution!** Do not breathe any of this gas, as it is a vigorous poison. What are the chief properties of this new gas? Compare it especially with the gas from which it was made. What do you consider the new gas to be? How formed? How distinguished from the first gas you made from carbon? This taking away of oxygen [or any similar substance] from another substance is called **reduction.** *Reduction* is the opposite of *oxidation*. Mention several cases of oxidation we have already had. Mention several of reduction, and tell in each of the latter by what means the reduction was accomplished.

Note. If the non-combustible oxide of carbon did not contain *at least two parts of oxygen* what would be left in the bags when the zinc had taken to itself one part of oxygen to form the zinc oxide that you saw in the tube? What *was* left in the bags? Are we not justified, then, in writing the non-combustible oxide as if it contained a double portion of oxygen? Let us hereafter call this oxide the *di*oxide of carbon.[1]

The chemical change brought about in this experiment may be represented by a diagram, thus:

$$\text{Zinc} + \begin{matrix}\text{carbon}\\\text{oxygen}\\\text{oxygen}\end{matrix} = \begin{matrix}\text{zinc}\\\text{oxygen}\end{matrix} + \begin{matrix}\text{carbon}\\\text{oxygen}\end{matrix}$$

---

[1] The old-fashioned name for this oxide is *carbonic acid*. It is not, however, an acid.

Experiment 20.

Oxidation of the Combustible Oxide of Carbon.

Have ready in a short, *e.g.*, 25^{cm}, piece of combustion tube[1] a small amount of the red oxide of mercury. Attach an empty rubber bag to one end of the tube and a bag containing the combustible oxide of carbon, made in Ex. 19, to the other end. Heat the oxide of mercury *gently* with a single Bunsen burner, and slowly pass the gas from end to end. Note the effect on the oxide of mercury. When no further effect is visible, cease passing the gas and test for the combustible oxide of carbon, then for the non-combustible. Use the flame test. How do you explain the change that has taken place? Draw a diagram that will show the change. Let us hereafter call this combustible oxide the *mon*oxide of carbon.[2]

Note. When charcoal instead of zinc is used in Ex. 19, there result from the one bag of the carbon dioxide two bags of the carbon monoxide. Explain this doubling of volume. If these two volumes of carbon monoxide should be passed over hot oxide of mercury how many volumes of carbon dioxide would result?

Experiment 21.

Action of Zinc on Sulphuric Acid.

Make an experiment parallel to Ex. 14, but use zinc, best in the granular form, where you there used iron.

[1] Hard glass tube, with a bore of 10–20^{mm}.

[2] Carbon monoxide is also called, correctly, carbonous oxide. Compare the names of sulphur*ous* and sulphur*ic* oxides and acids.

Be sure you omit no part of the experiment. Of what simple substances are the crystals composed? *Note.* This experiment teaches you one of the best ways known for making large quantities of a certain gas. What gas? Draw diagrams to show the changes. Compare the diagram of Ex. 14.

Experiment 22.

Action of Oxide of Zinc on Sulphuric Acid.

Proceed as in Ex. 21, but here use oxide of zinc. Be sure you note how the oxide behaves when put into pure water as well as when put into water and sulphuric acid. Why is hydrogen not given off in this experiment as in Ex. 21? State what has happened chemically in this [22d] experiment. Also state the **products** of the chemical change. What were the **factors** of the chemical change? What simple substances combined made each factor? What simple substances combined make each product? What is a chemical factor, — what a product?

Note that a diagram like the following will not only show the simple substances in both factors and products, but will express the chemical change that has taken place.

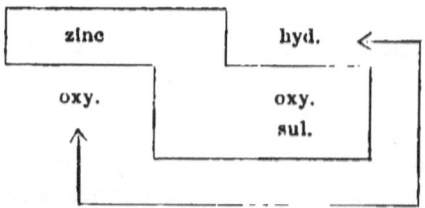

Experiment 23.

Sulphides.

A. **Mutual Action of Iron and Sulphur**, when warmed; or, **Formation of Sulphide of Iron.**

Take about one-half a cc of iron filings and an equal volume of flowers of sulphur. Mix well, then heat well in a large bulb tube [1] or in a tt. Break the tube and examine the substance formed. Let us call this **sulphide of iron**. Get its chief properties. Express the change thus: iron + sulphur = sulphide of iron; or thus: iron + sulphur = $\begin{bmatrix} \text{iron} \\ \text{sul.} \end{bmatrix}$.

B. **Action of Hydrogen on Warm Sulphur; or, Formation of Sulphide of Hydrogen.**

Generate hydrogen as follows: Place in a 250^{cc} flask about 100^{cc} of water and about 20^{cc} of sulphuric acid. Add about 30^g granular zinc. Have ready a tt, with about 10^g sulphur [best in the form of roll brimstone] in it; also a tube to connect the hydrogen flask with the tt, so that hydrogen can be conducted two thirds of the way down into the tt of sulphur. The sulphur tt should be fitted with a two-hole cork. Through one hole of this cork enters the hydrogen tube, which passes two thirds of the way down into the tt; through the second hole of the cork passes out an exit tube. The exit tube, made of hard glass, should start even with the lower surface of the cork, pass up through the cork, then be bent at a right angle. Toward its

---

[1] Same as Matrass of Ex. 5, A.

outer end the exit tube should be drawn down to a capillary tube and turned up at a right angle in the midst of this capillary part. The capillary part should have a total length of not less than 10^{cm}. **Caution!** Pass the hydrogen till safe to light it at the tip of the capillary tube. Prove that all is safe by the explosion tube, as in Ex. 9, C. Then boil the sulphur till the vapor of the sulphur nearly fills the tt. If the hydrogen is still burning, blow out the flame. Note the odor of the new gas coming. Let us call this new substance **sulphide of hydrogen.** Of what must sulphide of hydrogen be composed? Place a Bunsen burner flame under the exit tube just before it is narrowed to the capillary part. What is the deposit in the capillary part? What, then, must be going off? Light the jet and see if your answer to the last question is correct.

Express all the chemical actions in the form of equations in a manner similar to that indicated for Part A of this experiment.

C. Action of Sulphuric Acid on Sulphide of Iron.

State the simple substances that have gone to make up each of these compounds. Put in a 250^{cc} flask 50^{cc} water and 15^{cc} sulphuric acid. Warm somewhat. Then put in 25^g sulphide of iron. When the action begins catch in a *dry* fruit jar some of the gas formed. Catch by displacement. How do you catch by displacement? Do not warm after the sulphide of iron has been added. What is the gas formed? Test a jarful by the flame test. What are deposited on the sides of the jar? What is left in the jar after the fire? Tell by the odor. How must this have been formed? Get the properties of sulphide of hydrogen, particularly odor and solubility. This gas is often called **sulphuretted hydrogen.** In this experiment what has happened to the sulphide of iron and to the sulphuric acid? Answer this question after doing as follows: Transfer the contents of the flask to a porcelain evaporating dish. Evaporate with a very gentle heat till a little of the liquid taken out and cooled will deposit crystals. Immediately filter the contents of the dish. Cool the liquid which runs through and examine the crystals deposited. What are these crystals?

Draw a diagram which will show the simple substances in both the factors and the products, and will indicate the chief chemical change.

Experiment 24.

Copper.

A. **The Properties of Copper.**

Take some sheet copper and some copper wire. Get the chief properties of copper, as color, lustre, hardness, tenacity, *etc.*

B. **Oxidation of Copper.**

Take small pieces of copper wire. Proceed just as in Ex. 1, B, but use copper instead of iron. What shall we call the black coating formed on the bits of wire? Get the chief properties of this new substance.

Save some for C. How much oxygen gas was taken from the air in this experiment?

C. Reduction of the Black Oxide of Copper to Copper.

Take the black oxide of copper made in B. Place this in a piece of hard glass tube. The tube best be drawn out and turned up, as in the hydrogen sulphide experiment. Generate hydrogen as in Ex. 23. When it is desirable to generate a constant stream of hydrogen, it is well to have a thistle tube as well as a delivery tube passing through the cork of the generating flask. The lower end of the thistle tube should pass nearly to the bottom of the flask. Why? Through the thistle tube portions of a mixture of one volume of sulphuric acid to five of water can be poured from time to time as the action lessens. In this way no air is admitted, as there would be if the stopper should be taken out and the acid solution poured in. **Caution!** Keep all fire away. Why? Dry the hydrogen by passing it through a catch-bottle of sulphuric acid. Then pass the hydrogen through the hard glass tube and over the oxide of copper. **Test with explosion tube.** When safe, light the jet of hydrogen, and as the gas passes, gently heat the oxide of copper. Note the phenomenon. Explain. What becomes of the oxygen that was united with the copper forming the oxide of copper? Examine your apparatus and see if your answer to the last question is proved correct. What is left where copper oxide was? What is a *reduction?* What is *oxidation?* Contrast reduction and oxidation.

Draw a diagram to show the chief chemical change in this part of the experiment.

Experiment 25.

Magnesium.

A. Properties of Magnesium.

Take some magnesium in the form of ribbon and powder. Get the chief properties of magnesium. Note particularly the color, lustre, tenacity, brittleness, and weight.

B. Oxide of Magnesium.

Make this substance. Describe your method of preparation. Get the chief properties of **oxide of magnesium.** Save some of the oxide for C.

C. Action of Magnesium Oxide with Water.

Place some of the oxide from B on a piece of litmus paper that has been turned *slightly* red with an acid. Add a drop of water and note effect on the litmus. Treat, in a tt, some more of the oxide with water. Filter the contents of the tt. Evaporate the filtrate to see if any considerable amount of magnesium oxide dissolved or formed a soluble compound. In filtering, the liquid which filters through is called the **filtrate,** while the substance left on the paper is called the **precipitate.**

D. Reaction of Magnesium and Sulphuric Acid.

Note. When two substances mutually act, *i.e.*, act each on the other, a **reaction** is said to take place. Follow through the chemical change that takes place when magnesium and sulphuric acid are allowed to react with each other. What do you mean by reaction chemically? Describe this experiment in full. Make use of your notes on all work with zinc and with sulphuric acid. What do you call the substances that are the products in this case? What would have been the products if you had taken oxide of magnesium instead of magnesium? Draw diagrams to show the changes.

Note the tendency of magnesium, a metal, to push the hydrogen from the acid, and to take the place of the hydrogen it has driven out. Mention all other cases you have dealt with in which a metal has pushed the hydrogen from an acid. Mention two cases of this kind in which the metal, when it began to act with the acid, was itself already joined to another substance. Watch for the manifestation of this power of a metal in all your following work, and whenever you note it make a record of it in your note-book.

Experiment 26.

Calcium.

A. The Properties of Calcium.

Examine a small bit of this interesting simple substance, and note its chief properties, as color, lustre, hardness, and ease with which the air acts upon it.

B. Oxidation of Calcium.

Burn a small bit of calcium in a clean porcelain crucible and identify the oxide as the same substance as **quick lime**. Get the chief properties of **oxide of calcium**.

C. Reaction of Oxide of Calcium and Water.

Take a lump of quick lime weighing about 20^g. Put it in a porcelain evaporating dish. Blow on it, from the wash-bottle, water as long as the water is absorbed. Do not add any more water than this. Wait a few minutes for the change to take place. Note phenomena. Remember that several times we have made acids from oxides and water. With moist litmus paper, both blue and red, test the properties of the new substance here formed. Also test with turmeric [1] paper. Test the solubility of the new substance by shaking some in a tt with water, filtering, and evaporating the filtrate. Compare with the solubility of the similar compound made from oxide of magnesium and water. Let us call this new compound **hydroxide of calcium** or **hydrate of calcium**. Save about 10^g of it for E. What simple substances are in hydroxide of calcium? What is **slaked lime?**

D. Action of Calcium on Water.

Have ready a short narrow tt, say, two or three inches long and half an inch wide. Fit this tt with a one-hole rubber stopper and delivery tube. The delivery tube should be bent like an S, and its total length should not be more than 4 or 5 inches. Also have ready, instead of the pneumatic trough, a beaker nearly full of water. Let there be standing in the beaker an ordinary tt, inverted, and full of water. Fill the short tt to its stopper with water, and put a

---

[1] See Appendix C.

small bit of calcium in this water. At once insert the stopper and hang the apparatus, by means of the S tube, over the edge of the beaker so that the gas evolved shall go up and displace the water in the long tt. With a flaming match test the gas evolved. What is it? Whence came it? Evaporate the residue in the short tt and note that hydroxide of calcium has been formed. Explain its formation.

E. Reaction of Hydrate of Calcium and Sulphuric Acid.

What is another name for hydrate of calcium? Take about 10^g of powdered hydrate of calcium. Add about 30^{cc} of water. Stir till the hydrate of calcium is well mixed with the water. Note consistency. **Caution!** Protect the eyes from spattering, and add slowly, with constant stirring, about 5^{cc} of sulphuric acid. Stir for about two minutes. The **sulphate of calcium** forms at once, and "sets up" pasty or even hard. Compare its solubility with that of the sulphates of iron, zinc, and magnesium. Draw a diagram to show the change.

Sulphate of calcium occurs in nature both with and without water of crystallization. **Gypsum** is the kind that has the water. If gypsum be heated [110°–120° C.] the water passes off. The residue is called **plaster of Paris.** Put a bit of gypsum in a bulb-tube and heat. Record observation. Plaster of Paris has the power of again taking up water, and in so doing solidifies. Make a paste of plaster of Paris and water and watch it solidify.[1] Record observations.

[1] A cast may be made by pressing into the soft plaster a coin whose surface has been slightly greased and allowing the coin to remain till the plaster has "set."

F. Reaction of Hydrate of Calcium and Carbonic Acid.

Note. Recall the reactions of sulphuric acid and the various substances you have tried with that acid.

Here use an aqueous solution of hydrate of calcium, called **lime-water,** best made as follows. Take finely powdered hydrate of calcium and mix with cold water. Let settle for a few minutes. Pour off, and reject as much of the liquid as possible, as it contains a considerable amount of alkaline impurities dissolved by the water from the lime. Mix this washed hydrate of calcium with more cold water. Filter. Keep the filtrate in a well-corked bottle. Take about 10^{cc} of the hydrate of calcium solution and 20^{cc} of carbonic acid water.[1] Mix, and heat in a beaker until the new substance appears. Let settle, and **decant,** or filter. Decant means to pour off carefully, without filtering, a clear liquid from a precipitate that has settled from the liquid. Examine the new substance and recognize it as the same substance as powdered **marble** or **chalk.** Let us call it **carbonate of calcium.** Test its solubility in water. Why did it not appear [when the carbonic acid and hydrate of calcium were mixed] until you heated? In order to answer the last question proceed as follows. Put about 10^{cc} of lime-water in a tt. Add

---

[1] The instructor should prepare the carbonic acid water by passing carbon dioxide gas into a bottle half filled with cold water, and shaking till the water has absorbed all the gas it can, or, better, buy the water [soda water] from any soda-fountain keeper. Keep the carbonic acid water in a tightly-stoppered bottle in a cool place. A beer bottle with a rubber stopper that can be snapped down is good for keeping this water.

1^{cc}, only, of carbonic acid water. Note result, then add the other 19^{cc} of carbonic acid water and shake to mix. Water which has dissolved *sulphate* of calcium is called **permanently hard water.** Water [containing carbonic acid, *e.g.*, water which has made its way through swamps where there is decomposing vegetable matter], which has dissolved *carbonate* of calcium is called **temporarily hard water.** Why is one called permanently hard and the other temporarily? How may temporarily hard water be made soft? Hard water is said to "kill" soap. Prepare a solution of soap thus. Put about one gram of shavings from soap in about 10^{cc} of water in a tt. Shake till most of the soap has dissolved. Have ready two tts, in one of which are 10^{cc} of distilled water, and in the other an equal volume of hard water, *e.g.*, water that has been filtered from sulphate of calcium, or water that, by means of carbonic acid, has been made to take up carbonate of calcium. By means of a pipette, or long tube drawn out at the end to a wash-bottle tip, take up some of the clear soap solution and drop this, drop by drop, alternately in the tt of pure water and in the tt of hard water. Shake each tt after every two drops, and note in which tt frothing is produced the sooner. Note how many drops are required to produce frothing in each case. What inference do you draw in regard to the relative values of pure and hard waters for washing purposes? Define **stalactite.** Define **stalagmite.**

When oxide of zinc acted with sulphuric acid, and sulphate of zinc was formed, we concluded that the reason no hydrogen was given off [as there was when zinc and sulphuric acid were put together] was because the hydrogen joined the oxygen of the zinc oxide and formed water. Remember that hydrate of calcium is made of oxide of calcium and water. If a similar thing happens when sulphate of calcium is formed, what simple things must there be in sulphate of calcium? And if a similar thing happens when hydrate of calcium and carbonic acid react to form the carbonate of calcium, what simple things must there be in carbonate of calcium? As confirmation of your answer to the last question, pass the non-combustible oxide of carbon through a tt of hydrate of calcium in aqueous solution [lime-water], and note that the same carbonate of calcium is formed. Explain the formation, making use of a diagram.

Note. Lime-water is a good test for the presence of the non-combustible oxide of carbon because the two so readily react and form carbonate of calcium.

[1] Hold a tt inverted over the flame of a candle. Then add about 2^{cc} of lime-water. Shake, and note effect.

[2] Try the products of combustion from the flame of common house gas with lime-water. Also test the gas itself for oxide of carbon.

[3] Take a tt half full of lime-water and, by means of a glass tube reaching well into the lime-water, blow your breath several times through the liquid. Note effect, and explain. The non-combustible oxide of carbon, or carbon dioxide, is also called carbonic anhydride. What is the meaning of the term anhydride? Why applied to this oxide?

G. Analysis of Marble [calcic carbonate].

Note. We have already made a synthesis of marble. What is a **synthesis**? Mention several syntheses you have made. Explain in full the synthesis of marble.

Take a piece of large hard glass combustion tube about 15^{cm} long and closed at one end. [Best get a piece of tube about 30^{cm} long and make two 15^{cm} tubes by melting and pulling apart in the middle.] Be sure the end is good and stout, not cooled too quickly, and not ending in a lump of thick glass that is likely to crack off when the tube is again heated. Fit a cork and delivery tube to the open end. Put in the tube about 10^g of coarsely powdered marble. **Caution!** *Let the delivery tube dip only the least possible distance below the surface of the pneumatic trough, as deep dipping causes back pressure which is apt to blow out the softened glass.* Heat the marble with a blast-lamp. Apply the blast gently at first. Collect in tts the gas evolved. Test it. What is it? Prove that your inference is correct. [See note at end of *F*.] State how you have already proved of what this constituent of marble is composed. As it is impossible to make all the gas come off in a closed tube, *i.e.*, without a current to carry it away, take a small lump of marble, hold it with forceps, and heat in the blast-lamp flame till it glows brightly. Note the **lime** [or calcium] **light**. Examine the residue, and recognize it as quick lime, *i.e.*, oxide of calcium. How did you form oxide of calcium? [See *B.*]

Mix well one gram of powdered oxide of calcium with half a gram of powdered magnesium. Put the mixture in a small dipper. **Caution!** Keep the eyes away. Heat over a Bunsen burner flame. The magnesium takes the oxygen away from the calcium with great eagerness. Note phenomena. Have ready the short tt and fittings used in *D*. Put the contents of the dipper in the tt. Nearly fill the tt with water. Put in the stopper quickly. Catch the gas given off. What is it? Use the flame test. What does the evolution of this gas prove in regard to the chemical change? What happened to the magnesium? Arrange your analysis in the form of a table. What is an **analysis**?

II. Reaction of Marble and Sulphuric Acid.

State the simple substances that make marble. State the simple substances that make sulphuric acid. Take 10^s of powdered marble. Put them in a flask, add 30^{cc} of water and then 6^{cc} of sulphuric acid. Catch the gas evolved, and test it. What is it? How do you prove your answer correct?

Note how the new white substance formed collects around the marble, prevents the acid from coming in contact with the marble, and thus hinders the action. If the new white substance was readily soluble in the water, there would be no such hindrance. Many substances which are soluble in water are not soluble in sulphuric acid, hence in many cases, *e.g.*, in the preparation of hydrogen by means of zinc and sulphuric acid, and in the preparation of sulphide of hydrogen from sulphide of iron and sulphuric acid, it is necessary to use water to dissolve some product which otherwise would retard, if not actually prevent, the action.

Evaporate the residue from the flask and recognize it as sulphate of calcium. If there are bad fumes on evaporating, put the residue under the hood,[1] and evaporate there. Prove that the residue is not carbonate of calcium. How do you prove this? Explain the chemical change that has taken place. How can you detect a carbonate?

Experiment 27.

Sodium. [Latin name, Natrium.]

A. The Properties of Sodium.

Examine a small bit of sodium,[2] and get the chief properties, particularly color, lustre, hardness, and attraction for oxygen. **Caution!** The attraction of sodium for oxygen makes sodium a dangerous substance to handle. When experimenting with sodium keep it away from every substance that has oxygen, *e.g.*, water, and the moisture of your hand, even.

B. Oxidation of Sodium.

Let a bit of freshly cut sodium be exposed to the air for five seconds. What happens? What is the coating formed? Place a bit of sodium in a porcelain crucible and, protecting the eyes by a glass plate, warm. Note phenomenon. Save some of the resulting substance for *C.*

---

[1] See Appendix G.

[2] Sodium is a dangerous substance if handled carelessly, or allowed to get into water. It should be kept under kerosene, or some similar liquid, whenever not in use.

SODIUM WITH WATER.

C. Reaction of Oxide of Sodium and Water.

Treat the oxide of sodium made in *B* with a few drops of water. Protect the eyes as in *B*. Test the solution with test papers. Is it acid or the opposite? Compare with the action of other oxides and water. Filter, — if dirty, — and evaporate to dryness. Examine the new substance, and let us call it **hydrate** [or better, hydroxide] **of sodium.** Why call it hydroxide? Why hydrate? Rub a little sodium hydroxide between the fingers and note the greasy feeling. Put a little hydroxide of sodium on the bottom of a beaker and leave it exposed to the air for an hour. The peculiar property thus manifested is called **deliquescence.** What is deliquescence?

Note. Substances that have the opposite effect from acids on test papers are said to be **alkaline.**

D. Action of Sodium on Water.

Have ready, in a beaker, about 100^{cc} of cold water and a tt, upside down, filled with water. Make a little wire gauze net on the end of a wire. Wrap a bit of sodium, not larger than a small pea, in the gauze and plunge it below the water. Catch, in the inverted tt, the gas evolved. **Caution!** Keep the eyes well protected, for if the sodium gets out of the gauze an explosion is likely to follow. Why? Examine the gas caught. What is it? Filter the liquid left in the beaker, if any impurities from the gauze have got in. Evaporate to dryness in a porcelain dish. Test a bit of the substance with moist test papers, and by feeling. What is it? Explain its formation.

E. **Reaction of Sodium Hydroxide and Sulphuric Acid.**

Make two aqueous solutions as follows: [1] Take 100^{cc} of water and 5^{cc} of sulphuric acid. [2] Take 100^{cc} of water and 8^g of hydroxide of sodium. Have ready a porcelain evaporating dish. Fill about one fourth of the dish with the sulphuric acid solution, and add about as much hydroxide solution. Stir. Then add first one and then the other till drops taken out on a rod and touched to test papers show the contents of the dish to be **neutral,** *i.e.*, neither acid nor alkaline. Note, this process is called **neutralization.** Evaporate the liquid in the dish to crystallization. Examine the crystals, and get their properties. What are they? Compare Ex. 26, *E.* Compare also the formation of the products in Exs. 21, 22, 23, *C*, and 25, *D*. Explain the formation of the compound in this case. What simple things in this compound? Is any water of crystallization present? Test the answer to the last question by experiment.

F. **Reaction of Sodium Hydroxide and Carbonic Acid.**

Make an aqueous solution of sodium hydrate of the proportions 1^g hydrate and 20^{cc} water. Take about 10^{cc} of carbonic acid solution, and add the solution of hydroxide of sodium till the liquid is no longer acid. Evaporate to dryness. Examine the residue. What is it? It is composed of what simple substances? What is **sal soda?** Test sal soda for a carbonate. Leave a good clear crystal of sal soda exposed to the air for an hour or more. Note the phenomenon. The property that here manifests itself is called **efflores-**

cence. What is efflorescence? Compare with deliquescence.

G. Reaction of Hydrate of Sodium and the Non-Combustible Oxide of Carbon.

Take 0.5^g of hydrate of sodium. Dissolve it in a tt half full of water. Pass the dioxide of carbon through the liquid for about two minutes. Evaporate to dryness. Examine the residue. Dissolve the residue in 5^{cc} of water, and test for a carbonate. Explain the formation of a carbonate in this case.

H. Reaction of Carbonate of Sodium and Sulphuric Acid.

Parallel to *E*. State factors and products. Give results in full, explaining, particularly, the chemical change.

I. Reaction of Sodium and Mercury.

Caution! Caution! Caution!

Use extreme care and do not let particles fly in your eyes when the two substances join.

Take a bit of sodium and a globule of mercury. Support the cover to a porcelain crucible on a ring of the ring stand. Warm the cover. Pour on a globule of mercury about 2^{mm} in diameter. Then put a bit of sodium of about the same size near the mercury. Set the stand on the floor. Stand at a distance, and poke, with a long rod, the sodium into the mercury. Note phenomena. Examine the resulting product. A substance formed from mercury and another metal is called an **amalgam**. This then is sodium amalgam.

Throw the amalgam into a small dish of water. Note that the sodium acts as pure sodium, but less rapidly, and that [after a few hours], when the sodium is all gone [into what?], the original mercury remains. Give chief properties of **sodium amalgam.**

Review your work with sodium and its compounds, and make diagrams to express the chemical changes that you have brought about.

Experiment 28.

Chlorine.

Take a jar of chlorine.[1] Get the chief properties of chlorine, particularly its color and odor. Also note its action on moist litmus. **Caution!** Chlorine is very irritating to the mucous membranes. In getting the odor, breathe only the least possible amount. In large quantities chlorine is a violent poison.

Experiment 29.

Chlorides.

A. **Chloride of Hydrogen.**

Caution! Caution! Caution!

Chlorine and hydrogen unite with frightful violence. Always think what you are going to do with chlorine before you act. Never let hydrogen and chlorine mix

---

[1] Chlorine gas should be provided by the instructor and given, in tts and fruit jars, to the students as needed. For a method for preparing chlorine, see Appendix H.

accidentally. Never let a mixture of chlorine and hydrogen remain standing in the light or near a fire. A mixture of chlorine and hydrogen explodes spontaneously if left in the sunlight. Finally, — *use your utmost good judgment.*

Have ready a jar of chlorine gas and a flask that is delivering hydrogen gas through a rubber tube ending with a *hard glass* wash-bottle tip. When safe, light the stream of hydrogen gas, carefully raise the cover a little from the jar of chlorine, and slowly pass the hydrogen flame down into the chlorine. Note the change in the flame as chlorine, instead of oxygen, joins the hydrogen. Note the properties of the product of the union, particularly state and acidity. Chloride of hydrogen is commonly called **hydrochloric acid.**

B. Chloride of Sodium.

Have ready a tt of *dry* chlorine, and a piece of freshly-cut sodium. The piece of sodium should be a slice cut very thin, and large enough to cover about one quarter of the nail of your little finger. *Keep the tt well away from the eyes.* Drop the sodium in the tt. Cork, and let stand over night. *Spread out the new substance to air.* Get the properties. Try, by taste, to recognize it as common table salt. Do not taste if any unchanged sodium remains. Why? What simple substances form table salt? Let us call it **chloride of sodium.**

C. Preparation of Hydrochloric Acid on a Large Scale.

Take sulphuric acid and table salt. State the simple substances that make up each of these compounds.

Have ready a small flask with cork and delivery tube. Put about 10^g of the chloride of sodium in the flask. Add about 15^{cc} of sulphuric acid. Warm if necessary to start the action. Catch the gas by displacement. It will be found to be a heavy gas. Why not here use the water trough for catching? Fill at least three jars. Be sure the jars are *full*. Tell when full by the smell. What is this gas? How have you made it before? Now get the properties of this gas. Explain its formation from table salt and sulphuric acid. Make a diagram to show the simple substances, and their change of place. Compare the action of sulphuric acid in this case with its action on sulphide of iron. State the factors and the products in both cases. Save two jars of hydrochloric acid gas for subsequent use.

D. Solubility of Hydrochloric Acid.

Take a jar of hydrochloric acid gas [made in $\dot{C}$]. Put in about 20^{cc} of water. Snap on the cover. Shake. Open under water. Is the gas soluble? Test, with litmus, for the acid property of the liquid. For convenience in handling, hydrochloric acid is usually sold in aqueous solution. The crude hydrochloric acid of trade is called **muriatic acid.** It is yellow from impurities.

E. Reaction of Hydrochloric Acid and Marble.

Have ready a small flask fitted with a one-hole cork and delivery tube. Put a few lumps of marble in the flask, and cover them with aqueous hydrochloric acid. Catch the gas evolved and test it. What is it? After the acid has ceased acting, remove any marble that may be left, and evaporate the liquid to dryness. Of what is marble composed? What then must be the residue from evaporation? Note the chief properties of **chloride of calcium**. Leave a little exposed for an hour or two to the air of the laboratory. What property do you note? Why is chloride of calcium a good chemical with which to dry gases and other substances?

Had sal soda been used in this experiment instead of marble, what gas would have been given off, and what would the residue on evaporation have been?

Draw diagrams to show the reactions, both when marble was used and in a case in which sal soda is substituted.

Note. This experiment teaches the best way to prepare a certain gas on a large scale. What gas?

F. Action of Sodium on Hydrochloric Acid.

Have ready a jar of hydrochloric acid gas and some sodium amalgam. To insure success the jar must be *full* of the gas, and the sodium amalgam must have been recently made, and freshly broken into very small pieces.[1] The hydrochloric acid gas must have been dried by passing it through a catch-bottle of sulphuric acid, or there will be danger of an explosion. Why? Nor must there be any moisture in the jar at all. Be sure the washer to the jar is well greased.

Sodium amalgam is here used, instead of pure sodium, because the mercury modifies the action of the sodium and makes it controllable. That the mercury itself does not act, to any appreciable extent, on the acid, may

[1] For the preparation, on the large scale, of sodium amalgam see Appendix I.

be shown, after the experiment, by noting that the mercury, in liquid form, may be found in the trough.

To the jar of hydrochloric acid gas add about 25^g of sodium amalgam. Shake well. Note phenomenon. Open under water. At once note how much vacuum there was, at the same time letting the unused amalgam drop into the trough. [A zinc trough must not be used here because mercury amalgamates with zinc and makes it brittle.] Put the cover on at once, but do not seal, for, if any amalgam is left in the jar, there may be an explosion. Why? At once apply a flame to the gas remaining in the jar. What gas is this? Of what is hydrochloric acid made? What becomes of the chlorine in this experiment?

G. **Action of Sodium Hydroxide with Hydrochloric Acid.**

Dissolve about 5^g of sodium hydroxide in about 50^{cc} of water. Neutralize this solution with one of somewhat diluted hydrochloric acid. Evaporate till crystals begin to form. Examine the crystals and recognize them by their color, form, and taste as table salt. Explain their formation. What was the other product of the chemical change? Draw a diagram.

Experiment 30.

Potassium. [Latin name, Kalium.]

Note, as you work with potassium, how closely it resembles sodium, and how much like sodium compounds are the corresponding potassium compounds. **Caution!** Same as in the use of sodium.

POTASSIUM COMPOUNDS.

A. The Properties of Potassium.

Examine a small bit of potassium and get its chief properties; particularly, color, lustre, hardness, and affinity for oxygen. Note that it has metallic properties. Is it a metal?

B. Oxidation of Potassium.

Take a bit of potassium and place it on a crucible cover exposed to the air. Cut the potassium to expose more surface to the oxygen of the air. Note the rapidity of the oxidation.

C. Reaction of Oxide of Potassium and Water.

Parallel to Ex. 27, C.

D. Reaction of Potassium and Water.

Have ready a fruit jar with enough water in it to cover the bottom. Drop in a bit of potassium as large as a small pea. Instantly step back two yards or more. Note phenomena. Explain the action completely. Evaporate the liquid to dryness. Test a bit of the residue with moist test papers and by feeling. What is it? Explain its formation.

E. Action of Potassium on the Dioxide of Carbon.

Have ready a piece of hard glass tube about 10 inches long. Also a generator which is delivering *dry* carbonic dioxide. [See note at end of Experiment 29, *E.*] Put in the tube a piece of potassium as large as a small pea. Attach the tube to the generator, and clamp it at a convenient height for heating with a Bunsen burner. Pass the oxide of carbon until it will put out a match at the open end of the tube. Then warm the potassium. Note that it burns *at the expense of the oxygen of the carbonic dioxide.* Note the black particles of carbon that have lost their oxygen: note also the white powder. What is the latter? Answer this as follows: After the action has ceased and the tube is cool, wash out the tube into a beaker, *using as little water as possible.* Filter from the carbon. Test the filtrate for a carbonate. How test? Explain completely the formation of a carbonate.

F. Reaction of Potassium Hydroxide and Sulphuric Acid.

Parallel to Ex. 27, *E.* Examine the sulphate of potassium carefully, as you will be referred to this again. Note particularly the color, form, and taste of the crystals. Hold some in a Bunsen flame and note flame coloration. Heat some of the crystals in a porcelain dish and note that no water of crystallization is present. *Be sure you can recognize this substance again.*

Draw a diagram to show the chemical reaction. Also note the invariable tendency, again manifested in this change, of a metal to push the hydrogen from an acid. Mention all the other cases you can recall in which this tendency has been manifested.

Test the flame coloration of all the *potassium* compounds you can find. Note the flame coloration produced by table salt. Test the flame coloration of all the *sodium* compounds you can find.

Experiment 31.

Nitrogen.

We early learned that about one fifth the volume of the air is oxygen. Let us now examine the other constituent. Have ready a large cork with a little hollow dug out of its smaller end. Put about 0.3^g of phosphorus in the hollow of the cork. Float the cork on water in the pneumatic trough. Set fire to the phosphorus, and invert a jar of air over the burning phosphorus. What becomes of the oxygen of the air? What becomes of the oxide of phosphorus? Examine the remaining gas. Compare it with all the gases you have made. Let us call it **nitrogen**. Give the properties of nitrogen.

Experiment 32.

A Chemical Investigation.

There is often found as an efflorescence from the soil in hot countries, especially in Bengal, Egypt, Persia, and a few places in America, a white substance of a salt-like nature called nitre or saltpeter. This substance is used largely for making gunpowder, and often for making one of the most powerful acids known, — nitric acid. Let us direct an investigation to finding what simple substances are in nitre, and what ones are in nitric acid which is made from nitre. First let us prepare nitric acid.

A. **Preparation of Nitric Acid.**

Take nitre and sulphuric acid. Put in a small, tubulated, glass-stoppered glass retort, 30^g of nitre and 10^{cc} of sulphuric acid. Heat, gently at first, using a tripod and gauze. **Caution!** Do not put the hand where the molten mass or the acid could harm it, should there be an accident. Distil 10 or 15^{cc} of liquid, letting it drop so slowly that little or none of the vapor fails to condense. Get the properties of the liquid. It is so powerful that before testing it with test papers for acid properties, some of it should be diluted with several times its volume of water or it will destroy the paper itself. Let us call the liquid **nitric acid.** Have ready about 100^{cc} of water heated to 80° or 90° C. When the liquid in the retort has cooled somewhat, insert a funnel in the tubulature of the retort, and *cautiously* pour the hot water directly into the midst of the liquid. **Caution!** If the contents of the retort are *too hot*, there is danger that the hot water will be converted into steam so rapidly that an explosion may result, and if *too cold*, crystals may fix themselves so firmly to the sides of the retort that they cannot be removed without danger of breaking the glass. When the water is all added, stir the contents of the retort till the crystals are either dissolved or broken enough to be removed. Fill an evaporating dish with the liquid from the retort. Evaporate to crystallization. Examine the crystals. Test the flame coloration produced by these crystals and satisfy yourself by color, taste, form, *etc.*, that you now have the same substance made in Ex. 30, *F*. [It may be necessary, before you can get good crystals, to wash out the sulphuric and nitric acids that remain. Do this washing by putting the crystals on a filter, in a funnel, and letting a very small amount of cold water run through. Why not let much water run through? It would be well to recrystallize the substance from hot water to purify it.] What is the substance in hand? Of what simple substances is it formed? Remembering the effect of metals on the hydrogen of acids, what do you say here came from the sulphuric acid? What from the nitre? What then did the sulphuric acid lose, and what became of this that left the sulphuric acid? Review all the acids that you have made, *i.e.*, sulphurous, sulphuric, phosphoric, carbonic, hydrochloric, and note that all have hydrogen in them. What, then, do you say is one simple substance in nitre? What is one simple substance in nitric acid? As confirmation of your belief that there is hydrogen in nitric acid allow magnesium to act on some of the acid, as follows.

B. Action of Magnesium on Nitric Acid.

Have ready a tt fitted with a one-hole cork and delivery tube reaching to a pneumatic water trough. Put about 5^{cc} of nitric acid in the tt, and add about 10^{cc} of water. Put in a small strip of magnesium ribbon, and at once insert the cork and catch the gas in a tt at the trough. Apply a flame. What gas is present?

C. Action of Copper on Nitric Acid.

Take copper and nitric acid. Put about 50^{g} of copper clippings in a small flask. The flask should have a two-hole cork through which passes a funnel tube as well as a delivery tube. Add enough water to seal the funnel tube. Then add a mixture of 1 vol. of nitric acid to 1 vol. of water. Add this mixture a little at a time. Collect the gas over water. Catch three jars of it. Reject the first. Why? Get the properties of the gas from the second. Is it hydrogen? Note that on exposure to the air it oxidizes spontaneously, and the oxide formed is brown. In the third jar of gas burn about 0.3^g of phosphorus. Have the phosphorus well on fire before plunging into the jar. *Take all the usual precautions.* Note the formation of the white oxide of phosphorus. Therefore the substance under examination had oxygen in it. How much oxygen? Answer this by opening under water. From what sources may the oxygen have come? Test the residual gas. Use the flame test. Is it carbonic dioxide or nitrogen? Whence came this gas? What simple substances have we now proved to be in nitric acid?

D. Action of Carbon on Nitric Acid.

Take nitric acid and charcoal. Put in a Kjeldahl flask about 50^{cc} of nitric acid. Add three or four sticks of charcoal, each about as large as your little finger. Have a delivery tube leading to a catch-bottle containing water to catch any nitric acid that may pass over. Why must the acid be caught? Answer this question after the experiment is finished. Warm the nitric acid till gas passes off freely. After the gas has passed long enough to remove all air from the flask, catch-bottle, and tubes, collect some in tts and examine it. What is it? Pass some into a tt containing an aqueous solution of calcic hydroxide. What happens? Explain the formation of the gas in the Kjeldahl flask. What third simple substance must there be in nitric acid?

Note the vigor of nitric acid as follows. Put about 30^{cc} of nitric acid in a porcelain evaporating dish. Add a piece of cloth. Warm. Stir with a glass rod. Note effect on cloth. Stir the hot solution with a piece of wood. Note effect on wood. How do you explain the action on cloth, and on wood?

E. Reaction of Nitric Acid and Potassium Hydroxide.

Parallel to Ex. 30, *F*, except that you use nitric acid instead of sulphuric. Examine the resulting substance. Remembering the effect of metals on the hydrogen of acids, state what you think has been the change in this case, also what new substances have resulted, and of what simple substances each is composed. Let us call the white substance resulting **nitrate of potassium.**

Note. The substance resulting from the replacement of hydrogen in an acid by a metal is called a **salt,** *e.g.*, sulphates of calcium, magnesium, zinc, iron, *etc.*, carbonates of calcium, potassium, *etc.*, nitrates of sodium, potassium, *etc.*, are all salts to the chemist, as well as chloride of sodium. The last we have made by neutralizing hydrochloric acid with sodium hydroxide, as well as by the union of chlorine and sodium.

Identify, by color, form, taste, flame coloration, *etc.*, the nitrate of potassium formed in this experiment as nitre or saltpeter.

Finally answer our original questions. **What is nitre, and of what simple substances composed? What is nitric acid, and of what simple substances composed?**

To the Student. Up to this point you have worked out everything by yourself. You have yourself proved everything asked. However pleasant it may have been to discover all the truth for yourself, it is obviously unadvisable to proceed in this manner throughout your work in chemistry. Life is too short for you to *prove* to your own satisfaction all that has been discovered by all the workers in the field of chemistry in all the years. Now that you have completed a somewhat elaborate chemical investigation, and have learned how the pioneers of chemistry attack their problems, you are going to be asked to take much on faith, that is, you are going to be asked to believe many statements of facts without stopping to verify all.

Experiment 33.

Ammonia.

A. Preparation of Ammonia.

Take the colorless oxide of nitrogen of Ex. 32, *C*, and hydrogen. Prepare two jars of this colorless oxide of nitrogen as in Ex. 32, *C*. Also prepare five jars of hydrogen gas, free from air but not, necessarily, dry. Have ready in a pail, or tub, of water a bottle [large

AMMONIA. 69

enough to hold all the gases [1]], inverted, and full of water. Insert a large funnel in the neck of the bottle and pour the oxide of nitrogen and the hydrogen up into the large bottle.[2] Be careful, and do not spill any of the gases. The large bottle should be fitted with a two-hole rubber stopper. Through one hole let a piece of straight glass tube about 15^{cm} long pass, and through the other let a piece of glass tube bent at right angles pass just through the stopper. The straight piece should pass well into the bottle, and each should have a few cm of rubber tube attached to its outer end. Let each rubber tube have a pinch-cock.

Remove the funnel from the large bottle, insert the stopper with its tubes closed, and take the bottle from the pail. Connect the straight glass tube, by means of its rubber tube, to the water tap. Connect the bent glass tube to another tube containing platinum sponge or platinized asbestos. Let there be a catch-bottle containing water put just before the tube containing the platinum. Turn on the water, and force the gases through the water of the catch-bottle, and note the bubbles, in order to tell how fast the gases

---

[1] If you have not a bottle large enough, you can halve the amounts of the gases used, *i.e.*, take one jar of oxide of nitrogen, and two and one half jars of hydrogen.

[2] If you have not the pail and large funnel, you can proceed as follows. Take the large bottle and pour into it two jarfuls of water. Make a mark at the height to which the water reaches. Then pour in five more jarfuls, and make another mark at the height of the seven jarfuls of water. Next fill the remainder of the bottle with water, invert it on the bridge of the pneumatic trough, and pass the two gases, — first the oxide of nitrogen, then the hydrogen, — directly from the generating flasks into the bottle, stopping the flow of each when the water has fallen to the proper mark.

are passing. Gradually force the mixed gases out over the platinum. When the gases have driven all the air from the catch-bottle, *and not before*, heat the platinum. If the heat is applied before the air has gone, there is danger of an explosion in the catch-bottle. Why? Note the formation of water. From what is it made? Note the formation of a new gas. Note its odor and action on moist test papers. Of what is it composed? Let us call it **ammonia**. Pass some of the new gas into a tt containing a little water, and note that the water dissolves the gas. Try the action of the water solution on test papers. A water solution of the gas is sold in trade as "aqua ammonia."

B. Ammonia Fountain.

In order to show the great solubility of ammonia and the alkaline properties of aqua ammonia, make an ammonia fountain as follows. Have ready two 250^{cc} flasks,[1] each fitted with a good tight cork. Arrange a stand with a ring so that one flask can be supported, inverted, just above the other. Pass a glass tube through a hole in the cork of the lower flask so that it reaches to the bottom of the flask and projects about an inch above the cork. Through the cork in the upper flask pass a piece of tube drawn out to a fine point like the wash-bottle tip. The opening of the tip should have a diameter not less than 1^{mm} and not more than 2^{mm}. The tip should end at about the middle of the upper flask. Connect the glass tubes by a rubber connector. Bore a second hole in the cork of the lower

---

[1] Kjeldahl are best because they can stand more pressure without breaking than can ordinary flasks.

flask and pass a bit of glass tube bent at a right angle through it to admit air. Fill the lower flask one half or two thirds full of water. Add enough red [acid] solution of litmus to color the water distinctly red. Take the upper flask from its support, and fill it full of ammonia. Best prepare the ammonia by putting about 50^{cc} of commercial aqua ammonia in a flask and heating to drive off the gas. Conduct the gas into the flask by a rubber tube. Ammonia is a light gas compared with air. Get the flask *full* of ammonia gas. At once insert the cork and put the flask in position. Then connect with the lower flask. Blow, a little, in the air vent of the lower flask to start the fountain. Step back, as the flask may burst from the violence of the action. Note phenomena [physical and chemical]. Explain.

C. Salts of Ammonium.

What is a salt, chemically? How have we made salts?

An aqueous solution of ammonia behaves much like an aqueous solution of sodium [or potassium] hydroxide. It is alkaline to test papers, and will react with acids and form white salt-like substances ; in fact, it seems as though hydrogen and nitrogen together *act like a metal*. When thus acting together they are given the name of **ammonium**. Thus aqua ammonia is considered a solution of **ammonium hydroxide**.

1. Chloride of Ammonium.

Neutralize about 10^{cc} of aqua ammonia [diluted with 30^{cc} of water] with hydrochloric acid, — also diluted.

Evaporate to crystallization. Examine the product. What shall we call it? Put a little of this product in the bottom of a dry tt. Heat gently. Note *sublimate* on the sides of the tt. What is **sublimation**?

2. Sulphate of Ammonium.

Make this substance, and describe it. Will it sublime?

3. Nitrate of Ammonium.

Make this substance, and describe it.

Experiment 34.

Oxides of Nitrogen.

Note that we have already made two oxides of nitrogen: one a colorless gas, the other a brown gas. Review "A Chemical Investigation," Part C, and now state the proportion of oxygen to nitrogen [by volume] in the first, or colorless, oxide of nitrogen. What can you say in regard to the amount of oxygen in the second, or brown, oxide, compared with the amount in the first?

There are other compounds known which contain nitrogen and oxygen only. Of these perhaps the most interesting is the so-called **nitrous oxide** or **laughing gas**. Nitrous oxide is best prepared by decomposing ammonium nitrate by heat. Water and laughing gas are the products of this decomposition.

Take about 10^g of ammonium nitrate. Heat this gently in a large tt fitted with a one-hole cork and a delivery tube. Conduct the gases through three catch-bottles. Let the first catch-bottle contain no liquid, but be kept cold to condense the water formed in the decomposition of the nitrate. Let the second catch-bottle contain a little aqueous solution of hydroxide of sodium, to catch any acid-forming fumes. Let the third catch-bottle contain a little aqueous solution of sulphate of iron, to catch any of the [harmful] colorless nitr*ic* oxide that may be formed as a by-product. These by-products are most apt to be formed if the heat is applied too suddenly to the nitrate, or if the temperature is allowed to rise much above 170° C.

As nitrous oxide is soluble in cold water, the end of the delivery tube from the last catch-bottle should dip beneath the surface of some *warm water*, in a beaker or other vessel, and not into the cold water of a pneumatic trough. Catch the nitrous oxide in large tts. Get its properties. Particularly test it with a large glowing splinter of wood. Does it oxidize to a brown second oxide on exposure to the air, as did the first oxide of nitrogen you made? Inhale a little nitrous oxide.

Nitrous oxide has a greater proportion of nitrogen than does the first oxide of nitrogen you made. What proportion did you prove the first oxide has? That oxide is called **nitric oxide**. Compare the names sulphur*ous* and sulphur*ic*, as applied to oxides and acids of sulphur.

Note. At this point, before going on with Part II, the student should re-read carefully the Introduction to this book, and take note whether or not he has been following the suggestions there made.

PART II.

ADDITIONAL EXPERIMENTS.

PART II.—ADDITIONAL EXPERIMENTS.

Experiment 1.

Bromine.

EXAMINE a small amount of bromine and get its chief properties. Do not breathe in more than the least bit of bromine, for this substance is very bad for the mucous membranes.

Note particularly state, color, odor, volatility, and effect on moist test papers. To observe the properties it is best to take the bromine bottle to the hood, and with the window pulled so as to admit the arms only, pour a single drop of bromine into a tt and examine at leisure. On no account let a drop of bromine touch the flesh, as it makes a corrosive sore. Add about 5^{cc} of water to a tt full of bromine *vapors*. Shake. Note solubility. Warm, and note effect. Do this last experiment under a hood. Again prepare, in a tt, about 5^{cc} of bromine water. Add about 2^{cc} of ether. [**Caution!** Ether is dangerously inflammable. Have no fire near.] Shake. Note the solubility of bromine in ether as compared with its solubility in water. Note color of bromine solution in ether. Note also how slightly ether mixes with water. Is ether heavier or lighter than water? Try, in a similar way, the solubility of bromine in sulphide of carbon. [**Caution!**

Sulphide of carbon also is dangerously inflammable. Have no fire near.] Note the color of the solution of bromine in sulphide of carbon.

Experiment 2.

Bromides.

Hydrogen unites with bromine, but not so vigorously as with chlorine. By passing the vapors of bromine and hydrogen gas over hot platinum sponge the union may be made. The resulting hydrogen bromide is much like hydrochloric acid. It is also very soluble in water, and is commonly used in aqueous solution.

A. Properties of Hydrogen Bromide.

Take about 1^{cc} of aqueous hydrobromic acid in a tt, and get the properties both of the liquid and of the gaseous bromide of hydrogen that may be evolved by warming the liquid. Compare especially with hydrochloric acid.

B. Sodium Bromide.

Neutralize about 1^{cc} of hydrobromic acid solution with sodium hydroxide — or sodium carbonate — solution. Evaporate to dryness. Examine the product. Get the chief properties of the sodium bromide. Compare it with table salt, *e.g.*, in color, form, and taste.

C. Replacement of Bromine, in a Bromide, by Chlorine.

Make, in a tt, a solution of 10^{cc} of water and 0.1^g of bromide of potassium. Add about 5^{cc} of chlorine water, *i.e.*, water which has dissolved chlorine gas.[1] Add about 1^{cc} of sulphide of carbon. Why add the sulphide of carbon? Shake. Note effect. What has happened?

Experiment 3.

Iodine.

A. Properties of Iodine.

Examine a crystal of iodine and get its chief properties. Note particularly state, color, solubility, and color of vapor. To note the last, have ready a tt hot at the lower end. Drop in a crystal of iodine and examine the vapor.

B. Solubility of Iodine.

Get the relative solubility of iodine in the following solvents: — water, alcohol, ether, and sulphide of carbon. **Caution!** Remember that alcohol and ether, as well as sulphide of carbon, are dangerously inflammable. Have no fire near. Save, for *C*, the alcohol solution, which is called **tincture of iodine.**

C. Action of Iodine on the Skin.

Drop a small drop of iodine solution, saved from *B*, on the skin. Note effect.

D. Action of Iodine on Starch.

Prepare some starch paste as follows: Have ready in a porcelain evaporating dish 50^{cc} of boiling water.

---

[1] Chlorine water may be made by allowing chlorine gas to bubble up through cold water contained in a catch-bottle, flask, or other vessel.

Rub in a mortar 1^g of starch and a few drops of water to the consistency of cream. Stir the starch into the boiling water and set aside.

Put in a tt about 1^{cc} of water, add two or three drops of starch paste, then add [after shaking till the paste is well mixed with the water] a single drop of an aqueous solution, or a minute crystal, of iodine. Heat the solution and note the effect. Cool, and note again. Dip a strip of filter paper in the starch paste and suspend it across the mouth of a tt containing a small crystal of iodine in the bottom. Note effect. What does this show?

Experiment 4.
Iodides.

Hydrogen unites with iodine, but not readily. By using platinum sponge and iodine vapors the union may, with difficulty, be made. Hydrogen iodide, or hydriodic acid, as it is usually called, resembles hydrochloric and hydrobromic acids. It forms iodides similar to bromides and chlorides.

A. The Properties of Potassic Iodide.

Examine this substance. Note its chief properties. State what relation it has to hydriodic acid, to bromide of potassium, to chloride of potassium.

B. Replacement of Iodine by Chlorine.

Make a solution, in a tt, of 10^{cc} of water and 0.1^g of iodide of potassium. Add about 5^{cc} of chlorine water. Then about 1^{cc} of sulphide of carbon. Shake. Note effect.

FLUORINE AND FLUORIDES.

Stir a crystal of potassic iodide into a starch solution. Note effect. Add a few drops of chlorine water. Note effect. Explain.

C. Will Bromine Displace Iodine?

Perform an experiment to prove which has the stronger attraction for potassium — iodine or bromine. Record details of the experiment, and the conclusion reached.

Experiment 5.

Fluorine and Fluorides.

We have studied iodine, bromine, chlorine, and the corresponding iodides, bromides, and chlorides. It has long been known that there must be a fourth member of this halogen[1] group, because salts were known — called fluorides — similar to the bromides, chlorides, and iodides. The simple substance itself has recently been prepared and is called **fluorine.** Fluorine is much like chlorine, but acts more energetically.

A. The Properties of Calcic Fluoride.

Examine some **fluoride of calcium** and get its chief properties.

B. Preparation of Fluoride of Hydrogen.

Compare the "preparation of hydrochloric acid on a large scale," Ex. 29, C. Put about 1^g calcic fluoride

---

[1] Halogen means salt-making. Note that chlorine makes common salt, and bromine and iodine similar salts.

in a tt. Add enough sulphuric acid to make a paste. Warm gently. Smell *cautiously*. *A small amount of this gas can seriously injure the lungs.* Hold a glass rod wet with water in the mouth of the tt. Get the chief properties of hydrofluoric acid.

C. Etching of Glass by Hydrofluoric Acid.

Hydrogen fluoride is an extremely corrosive substance. All metals, except lead, gold, and platinum, act with it to form metallic fluorides. Even glass is eaten away by hydrogen fluoride, with the formation of silicon fluoride, the silicon coming from the sand which was used to make the glass. Hence hydrofluoric acid cannot be kept in a glass bottle. It can be kept in lead, platinum, or rubber.

Warm a bit of glass, *e.g.*, the bottom of a beaker or of a crystal pan, and drop on it a little wax or candle grease. Spread the grease thin, and let it harden by cooling. Then trace some figure or name in the grease, cutting through to the glass. Generate hydrogen fluoride again, as in *B*, using for a generating vessel a small beaker, a large tt, or, still better, a lead dish. Expose the prepared glass to the action of the vapors of hydrogen fluoride for a few minutes. Do not melt the grease while the vapors are acting. Then warm the glass, to melt the grease, and wipe off the grease with filter paper. Examine the surface of the glass. Breathe on it. What do you note? Explain.

Experiment 6.

Arsenic and its Compounds.

Caution! In all your work with arsenic and its compounds **use extreme care,** and do not get poisoned.

A. The Properties of the Simple Substance — Arsenic.

Examine a small bit of arsenic and note its chief properties. Keep it well away from the mouth. Do not test its *chemical* properties without the advice of the instructor. Put a piece, not larger than a common pin's head, in a hard glass bulb tube, clean and dry, *with a long stem.* Heat gently. Note effect. Be sure you put the tube when used [and all remnants of arsenic and arsenic compounds] in a small waste-box provided for the purpose.[1]

B. Oxidation of Arsenic.

Do this under a hood with window pulled down. Burn, on the cover of a porcelain crucible, a bit of arsenic, not larger than a pin's head. Note the oxide formed. This white oxide is the "white arsenic," or, more commonly, "arsenic," of the apothecaries. Describe it.

Again burn arsenic [the smallest possible amount] in order to get the peculiar odor of the oxide. Note that it has a garlic odor, and reminds one of onions.

---

[1] These arsenic residues should not be thrown with the other laboratory waste, but should be guarded from contact with chemicals.

C. Reduction of the Oxide of Arsenic.

Carbon has the power of removing the oxygen from this oxide. Compare the removal of oxygen from the non-combustible oxide of carbon by means of zinc. Fill the bulb of a hard glass matrass half full of a mixture of equal parts of powdered charcoal and oxide of arsenic. Warm gently. Note **arsenic mirror.** Is all the oxygen taken from the arsenic, or is a lower oxide left as in the case of the reduction of the dioxide of carbon?

D. Arsenide of Hydrogen.

Caution! Arsenide of hydrogen is one of the most poisonous substances known. Use every care.

Arsenic, like sulphur, forms a gaseous compound with hydrogen, similar to the sulphide of hydrogen. The arsenic compound is called **arsenide of hydrogen,** or **arsine,**[1] and is characterized by its frightfully poisonous nature. Take a small, *e.g.*, a 2 oz. or a 4 oz. salt-mouth bottle. Fit with a two-hole rubber stopper. Pass a funnel-tube through one hole and nearly to the bottom of the bottle. Through the other hole pass a short right-angled piece of glass tube. The inner end of this glass tube should be flush with the smaller end of the rubber stopper. With the right-angled tube, connect a tube containing chloride of calcium in small lumps. What is a chief property of chloride of calcium? Connect with the chloride of calcium tube a piece of hard glass tube drawn to a pin-hole bore and turned up at a right-angle, similar to the tube used

---

[1] Also called arseniuretted hydrogen.

for the production of hydrogen sulphide, Ex. 23, *B*. Put in the bottle about 5^g of c.p.[1] zinc, and a few cc of c.p. sulphuric acid, diluted, *e.g.*, 1 vol. acid to 5 vols. of water. Let the hydrogen evolved pass through the tubes till all the air is removed and till the gas can be lighted [by the explosion tube] when it issues from the turned-up small glass tube. When the hydrogen is burning well, place the flame of a Bunsen burner under the hard glass tube just before the point where it narrows to a pin-hole size. *Do not remove this flame till the experiment is finished.* Then add, through the funnel-tube, to the bottle five drops, and no more, of an arsenical solution. Use the arsenical solution especially prepared by the instructor for this purpose and no other solution.[2] Wash the whole of the solution used down the funnel tube into the bottle by means of pure water from the wash-bottle. Continue the heating for 30 minutes at least. The hydrogen of the sulphuric acid joins the arsenic and the arsenide of hydrogen passes off. Compare the formation of sulphide of hydrogen. The gas is decomposed as it passes through the hot tube, and the hydrogen passes on and burns, while the arsenic is left, as a mirror, in the contracted tube. Compare the action of heat on sulphide of hydrogen. Keep the arsenic mirror for future work.

E. Detection of Arsenic.

D gives us a method for detecting arsenic. It is always necessary first to make sure that the arsenic

---

[1] In chemistry, c.p. stands for chemically pure.
[2] See Appendix J.

is in such a form that the hydrogen can join it and form the arsenide of hydrogen. Therefore the substance which you suspect contains arsenic should first be treated with sulphuric acid [or, in the case of a piece of cloth, carpet, or other substance likely to contain wool, with nitric and sulphuric acids].

Arrange the apparatus as in D. Start the hydrogen, and when safe to light the jet do so, and put the Bunsen burner in place. Allow the hydrogen alone to pass for 15 minutes with the tube at a dull red. This preliminary test is to make sure that there is no arsenic in the zinc or the acid used, and none left from a previous experiment.

Cut from a piece of wall paper, or any similar suspected material supposed to contain arsenic, one square decimeter. Be sure you include all the colors of the figures, if the material is figured. Tear the substance in small bits and put these in a clean and dry porcelain evaporating dish. Add a little c.p. sulphuric acid. Warm gently, and stir with a glass rod till solution takes place. The solution will not be transparent, and is generally black. Do not use enough acid to make the solution really liquid — it should be somewhat pasty.[1] Add about 20^{cc} of cold water, or an equal weight of ice or snow, to dilute the strong acid. When cool, filter. The arsenical solution runs through. If there is any precipitate, wash this a little and add the washings to the filtrate. If the apparatus has shown

---

[1] If the substance contains wool its sulphur must now be oxidized. Add to the sulphuric acid solution about 5^{cc} of c.p. nitric acid. Evaporate till white fumes appear. Repeat this treatment. Then dilute with water, again evaporate till white fumes appear, and go on as above.

no arsenic after the preliminary test of 15 minutes, add the solution to be tested through the funnel. Be careful and do not let many bubbles of air go down the funnel-tube with the solution. If arsenic now appears the experiment should be continued for 30 minutes in order to catch all the arsenic as a mirror. **Caution!** If arsenic appears, do not break any joints of the apparatus, nor remove the Bunsen burner, for 30 minutes, or there will be danger of being poisoned by the escaping arsenide of hydrogen. It is also well to keep the little jet of hydrogen burning all the time to decompose any hydrogen arsenide that may escape the decomposition above the Bunsen burner. If the wind blows the flame of the Bunsen burner, protect it with a book, or something else, or the arsine may escape undecomposed.

Experiment 7.

Antimony. [Stibium.]

In doing this experiment note the similarity between antimony and arsenic.

A. **The Properties of Antimony.**

Get the chief properties of antimony. Try subliming some in a small matrass.

B. **Oxidation of Antimony.**

To oxidize it well, heat it on a crucible cover before a small mouth blow-pipe. Place on the cover a piece of antimony as large as a pin's head. Using a Bunsen burner, direct the blow-pipe flame [1] against the antimony. Note the fumes of oxide that rise. Suddenly stop the blast, and note the globule of molten antimony as it becomes coated with oxide. Drop a small molten globule of antimony from some height down on a piece of paper whose edges are turned up [to prevent the antimony running off], and note effect.

C. Chloride of Antimony.

Take a jar of chlorine. Sift in a little antimony, which has been ground, in a mortar, to the finest possible powder. Note the phenomena, and examine the compound formed. Review the formation of table salt.

D. Hydrogen Antimonide.

Treat a solution of some antimony compound [2] in the apparatus used in Ex. 6, *D*. Hydrogen joins the antimony, and gaseous antimonide of hydrogen results. Examine the antimony mirror, and compare it with the arsenic mirror, particularly as to position in the tube, color, and lustre. Heat each gently, and see which sublimes easier. Prepare a solution of bleaching lime [about 1 part lime to 10 of water] and dip, alternately, the mirrors in this solution. Which dissolves the easier?

---

[1] For use of blow-pipe, see Appendix K.
[2] See Appendix J, 2.

BISMUTH.

E. **A Chemical Examination.**

Examine two pieces of paper,[1] one containing arsenic, the other antimony. Compare the two mirrors which you get, and determine which is arsenic and which is antimony.

Experiment 8.

Bismuth.

A. **The Properties of Bismuth.**

Examine a small piece of the metal and get its chief properties.

B. **Nitrate of Bismuth.**

Prepare nitrate of bismuth by putting a lump of bismuth in a small amount of nitric acid. Heat somewhat. Add to this nitrate about 1^{cc} of water. Note solubility. Then add the solution to 100^{cc} of water and note the formation of a basic nitrate, insoluble in water, which settles out as a light milky precipitate.

Experiment 9.

Tin. [Stannum.]

A. **The Properties of Tin.**

Examine tin in various forms, *e.g.*, bar, granular, foil, and on iron as "tin plate." Get the chief properties of tin, noting particularly its color, lustre, hardness,

[1] See Appendix L.

"cry," and whether it is much or little affected by water and the air. To note its cry, bend a bar or pinch it between the teeth.

B. Oxidation of Tin.

Oxidize granular tin in a small Hessian crucible over a blast-lamp, or in a Fletcher furnace. Stir often. Examine the oxide and get its chief properties.

C. Crystalline Structure.

Take a piece of "scrap tin plate," *i.e.*, a bit of tinned iron that has been cut off in the making of tin ware. Heat this over the Bunsen burner flame till the tin begins to run. Plunge into cold water suddenly. Remove the superficial oxidation by rubbing the surface with a bit of filter paper wet in nitric and hydrochloric acids. Remove the acids by rubbing with a weak solution of sodium hydroxide. Wash off the sodium hydroxide with water. Examine the crystalline structure presented by the tin.

D. Action of Strong Acids on Tin.

1. Treat tin, in a tt, with hydrochloric acid — cold and hot.
2. Treat tin, in a tt, with sulphuric acid — cold and hot.
3. Treat tin, in a tt, with nitric acid — cold and hot.

Note effect in each case. Save substances formed.

LEAD.

E. **Replacement of Tin by Zinc.**

Use the chloride of tin made in *D.* Make a strong solution of chloride of tin. Put this in a large tt. Clean a narrow strip of zinc and insert it in the solution. Note effect. Explain.

Experiment 10.

Lead. [Plumbum.]

A. **The Properties of Lead.**

Examine a piece of lead and note the chief properties.

B. **Oxidation of Lead.**

Put some lead in a Hessian crucible and heat it over the blast-lamp. Stir to admit air. Note that the lead forms several oxides of different colors. Do not let the heat become very great. Save some oxide.

C. **Action of Water on Oxide of Lead.**

In a mortar grind a little of the oxide of lead, from *B*, with about 5^{cc} of water. Filter and see if anything has gone into solution. Test the liquid also with test papers. State what you can about this experiment from a chemical point of view, — from a sanitary point of view, *e.g.*, in reference to the use of lead pipes for conveying drinking water, inasmuch as soluble compounds of lead are poisons.

D. **Action of Acids on Lead.**

Similar to 9, *D*. Record results in tabular form.

E. Replacement of Lead by Zinc.

Similar to 9, *E*, except a solution of lead nitrate best be used. Note the formation of a "lead tree." Explain.

F. Lead Chloride.

Treat a strong solution of 5^g of lead nitrate with an excess of hydrochloric acid. Explain the replacement that takes place. Dilute a little and filter. While the precipitate of lead chloride is still on the filter, wash it a little with a stream from the wash-bottle. Get its chief properties. Put the precipitate of lead chloride in a tt. Add three times its volume of cold water. Warm. Note solubility in hot water. Let the solution cool, and note the formation of crystals.

G. Lead Sulphate.

Compare the results of *D* for the action of sulphuric acid on lead itself. Now prepare lead sulphate by adding sulphuric acid to a solution of lead nitrate. Wash the precipitate, as in *F*, and get its chief properties. Note that this is a roundabout way for preparing a salt when we cannot readily get it by treating the metal with the acid. Such roundabout methods are often used by the chemist.

H. Plumbers' Solder.

Melt, in a Hessian crucible, equal parts of tin and lead. The resulting alloy is common **solder**. Note how much more easily solder may be melted than either lead or tin. Two or more metals thus fused together form what is called an **alloy**.

I. Fusible Alloy.

Take 15^g of bismuth, 8^g of tin, and 8^g of lead. Put each under boiling water, and note that no one of them melts. *Dry* the metals, and fuse the three together in an iron spoon or a crucible; cool, and place the alloy thus produced in boiling water. Note effect. While still melted, pour the alloy into a narrow, thin-walled tt. Let cool. Note effect. Explain.

Experiment 11.

Silver. [Argentum.]

A. The Properties of Silver.

Examine a small piece of silver and note its chief properties, particularly the effect of air and water on it.

B. Oxidation of Silver.

Attempt to form an oxide of silver in all the ways you know for oxidation.

C. Action of Acids on Silver.

Try your strongest acids — sulphuric, hydrochloric, and nitric — "in the cold"[1] and when hot. Evaporate some of the acid after each attempt, and compare the

---
[1] This expression means at ordinary temperature.

amounts of residue. Try dilute and strong acids. Save any salts found. Make a carefully prepared table of results.

D. Replacement of Silver by Copper.

Take 0.5^g of silver nitrate, dissolve it in 10^{cc} of water. Add a freshly cleaned strip of copper. Note effect. Spread the silver on a hard surface and rub it with some hard instrument, as a knife blade, to restore its lustre.

E. Replacement of Silver by Calcium, Sodium, and Potassium.

1. Replacement of silver by calcium, and the formation of silver chloride.

Put a small crystal of nitrate of silver in about 10^{cc} of water. Shake till solution takes place. Have ready a solution of a small lump of calcic chloride in about 10^{cc} of water. Mix the solutions. Filter. Examine the precipitate of silver chloride. Expose some to light — sunlight is best. Note effect. Explain the changes. Name the factors and the products.[1]

2. Replacement of silver by sodium, and the formation of silver bromide.

Proceed as in 1, but use sodic bromide for calcic chloride.

3. Replacement of silver by potassium, and the formation of silver iodide.

---

[1] After every experiment in which silver or silver nitrate is used, any residues (solid or liquid) containing silver should be saved. When a considerable quantity of these residues has been obtained, the instructor should regain the silver from them. See *II*, Method II.

PURIFICATION OF SILVER. 95

Proceed as in 1, but use potassic iodide for calcic chloride.

F. Sulphide of Silver.

Dissolve a small crystal of nitrate of silver in a tt half full of water. Generate sulphide of hydrogen [from sulphide of iron and dilute sulphuric acid in a flask or in a tt], and pass the gas through the nitrate solution in the tt. Examine the sulphide of silver formed. Explain the replacement that has caused the formation of silver sulphide. What is left in solution? Hold a silver coin for a moment in a stream of sulphuretted hydrogen. Note effect.

G. Oxide of Silver.

Add about 0.5^g of nitrate of silver to half a tt of water. Then add a few drops of a strong solution of sodium hydroxide. Examine the silver oxide formed. Explain the chemical changes.

II. Purification of Silver.

METHOD I. BY REPLACEMENT WITH COPPER.

Dissolve a silver coin, *e.g.*, a dime, by heat, in dilute nitric acid. There is copper in silver coins. What two salts then are in the nitric acid solution? If there seems much nitric acid left it should be mostly evaporated off. Insert a clean strip of copper and set aside for some time. Explain the change. Collect the silver deposited. Throw it on a filter, and wash thoroughly with water from the wash-bottle. Why wash? Evaporate the filtrate and washings to dryness. Examine the substance left. What is it?

Method II. By Reducing a Chloride.

Dissolve a coin, as above, in nitric acid. Add hydrochloric acid [or a solution of table salt] as long as a precipitate is formed. Explain the chemical changes. Filter. Wash the precipitate thoroughly. Why? Dry the precipitate.[1] Remove the chloride of silver from the paper, and put the chloride in the midst of a piece of combustion tube. Generate hydrogen gas. *Take every precaution.* Fit two corks to the combustion tube, and pass hydrogen gas over the chloride and out through an exit tube. Light the hydrogen by the explosion tube *and in no other way*. When the hydrogen is burning, *and not before*, heat the chloride well. Note the effect on the hydrogen flame. Explain. When all the chloride is reduced, examine the silver left. What is reduction?

Experiment 12.

Gold. [Aurum.]

A. The Properties of Gold.

Examine a small piece of gold [leaf or foil will do] and note its chief properties, particularly its color, hardness, and malleability.

B. Action of Acids on Gold.

Try your strongest acids on gold. Gold is called the "king of metals" because of its resistance to the action of acids.

[1] See Appendix M.

C. Chloride of Gold.

By treating hydrochloric acid with nitric acid, chlorine may be set free. Explain this. At the moment chlorine is set free, or, as it is called, at the moment of its birth [commonly "in statu nascendi"], chlorine has unusual power of uniting with other things. Explain this great power of **nascent chlorine.** Nascent chlorine can, with success, attack gold and produce gold chloride.

Put in a watch-glass a drop or two of hydrochloric acid and a drop of nitric acid. Warm, and drop in about 1 sq. cm. of gold foil. Note effect. Carefully evaporate the solution and examine the gold chloride formed. The mixture of hydrochloric and nitric acids is called **aqua regia,** or royal water, as it attacks gold, the king of metals. Silver and gold are called the **noble metals** because they do not tarnish in ordinary air.

D. Gold Amalgam.

Put a small drop of mercury on a watch-glass. Spread over it a piece of gold leaf as large as a sq. cm. Note effect. Why should a gold ring be kept out of mercury?

E. Color of Gold.

Examine a piece of gold leaf, or foil, as it lies with the light *reflected* from it. Hold a piece of gold leaf up to the light and note the color as the light is transmitted *through it.* Best lay the leaf on a plate of glass for convenience in handling.

Experiment 13.

Platinum.

A. The Properties of Platinum.

Examine a small bit of platinum and get the chief properties, particularly color and fusibility. Why is platinum an excellent substance from which to make crucibles?

B. Action of Acids on Platinum.

Try your strongest acids on platinum. Also aqua regia. Save any salts formed.

C. Action of Other Chemicals, besides Acids, on Platinum.

Try other chemicals [than those of *B*], *e.g.*, alkalies, salts, *etc.* Why is platinum an excellent substance for chemical utensils?

D. Action of Metals with Platinum.

Heat, in a clean crucible, a bit of platinum with a bit of lead. Note the formation of a fusible alloy. Why should not metals be heated in platinum dishes?

E. Platinum Sponge.

Put about 2^{cc} of a solution of chloride of ammonium in a tt. Render acid with hydrogen chloride solution. Add, drop by drop, the chloride of platinum solution made in *B*. Note the formation of a yellow precipitate. Collect about 1^{cc} of this — having made sure that the ammonium chloride is in excess, and not the platinic chloride. Separate the precipitate, by decantation, from its liquor. Dry the precipitate a little by very gentle heat. When it is only slightly moist, put it in a bit of platinum foil, made into a little cup, and heat to redness in a small Bunsen flame as long as fumes are given off. All the simple substances, except the platinum, will go off. What then go off? Examine the **platinum sponge** formed. It is *metallic*. Direct a small jet of house gas [better, hydrogen] against a surface of freshly-made platinum sponge. Note effect.

Experiment 14.

Aluminum.

A. The Properties of Aluminum.

Examine the metal — ingot, sheet, or wire form — and note the chief properties, as color, lustre, hardness, and, approximately, the specific gravity.

B. Oxidation of Aluminum.

Try to oxidize aluminum.

C. Action of Acids on Aluminum.

Try all the acids you can — diluted and strong — hot and cold. Results in a table.

D. Sulphate of Aluminum.

Prepare the sulphate and crystallize it.

E. Alum.

Make, in a tt, a saturated solution of sulphate of aluminum. In a second tt make a saturated solution of sulphate of potassium. Take 10^{cc} of each of these solutions. Mix. Shake. Evaporate about one third the water. Let crystallize. Examine the crystals under a microscope, comparing them with the original crystals of sulphate of aluminum and sulphate of potassium. Recrystallize the alum from hot water. **Nurse**[1] **a good crystal.** Compare the solubility of the alum with that of the original sulphates. What is an alum?

[1] See Appendix N.

PART III.

HISTORY AND DEVELOPMENT

OF THE

LAWS AND THEORIES OF CHEMISTRY.

PART III.

LAWS AND THEORIES OF CHEMISTRY.

CHAPTER I.

INTRODUCTION.

CHEMISTRY may be defined as that department of human knowledge which has to do with those phenomena that result from changes of substance.

An examination of the many changes to which matter is subject shows that these changes fall into two groups — there are changes in which the composition of the substance is not altered, and others in which there is a *change of substance*. The first are called *physical changes*, the second *chemical*. As examples of physical changes may be given: the formation of the solid, ice, from the liquid, water; the glowing of platinum wire when the electric current is passed through it; the dissolving of sugar when stirred into water. In no one of these cases is it thought there is any change of substance, *i.e.*, the ice is oxide of hydrogen just as much as is the water, the platinum is still platinum, and the sugar remains sugar. But when the electric current was passed through water we found that the water disappeared, and two gases of unlike properties were evolved; when platinum was treated with aqua regia, in Ex. 13, *B*, Part II, a chloride not at all like the metal platinum resulted; when sugar is dropped on a hot stove a black charcoal is left, while water vapor passes off into the air. These last are **chemical changes** because in every one there has been a change of substance.

When a rod of iron becomes heated and glows with a red color, is the change a physical one, or is it chemical? When iron becomes magnetized, and is capable of attracting to itself other pieces of iron, to which class does the change belong? When iron burns and the black oxide is left, when it is acted on by moist air and the red rust is left, or when it is acted upon by sulphuric acid and the green sulphate is left, what are the changes, chemical or physical? Give the reasons for your answers. When glass is softened by heat is the change chemical or physical? When the iron filings were heated in contact with air was the blackening of their surface caused by a physical or a chemical change? When salt stirred into water disappears is the change physical or chemical? Answer this last question after doing the following experiment.

Experiment 1.

Two Kinds of Changes.

1. Dissolve about one gram of table salt in about 10^{cc} of water. Evaporate the water, and examine the residue as to color, taste, *etc.* Note that it is the same

salt that was taken, *i.e.*, there has been no change of the substance, therefore the solution is called a physical change.

2. Warm, in a tt, with a little sulphuric acid and about 10^{cc} of water, some iron filings till solution takes place. Evaporate the liquid, and examine the residue. What kind of a change has taken place? How do you know? Which dissolved — the iron, or the sulphate of iron which resulted from the chemical change? Would it be right, as is frequently done, to call this a chemical solution of the iron?

3. Have ready a porcelain evaporating dish containing 15–20^{cc} of water. Take about 5^g of dry powdered carbonate of sodium and stir this into the water till a clear solution results. Evaporate to dryness. Examine the residue. Compare it with carbonate of sodium in color, form,[1] and taste. What is it?

Next put 15–20^{cc} of hydrochloric acid solution in a porcelain dish. Add about 5^g of carbonate of sodium and stir till a clear solution results. Evaporate to dryness. Examine the residue. Compare it with the original carbonate of sodium in color, form,[1] and taste. What is it? What kind of a change took place when water alone was used? When hydrochloric acid was used? What went into the solution in the first place? What in the second? How do you distinguish between a simple **physical solution** and a **chemical solution**?

The loss or gain of water sometimes causes a change. Recall the change that followed when, in Ex. 14,

[1] The form can best be observed under a microscope of medium power.

Part I, the water of crystallization was driven out by heat from the green crystals of sulphate of iron; also the changes when, in Ex. 27, Part I, sal soda effloresced, and when [in the same experiment] sodium hydrate deliquesced.

Experiment 2.

Changes caused by Water of Crystallization.

Put a small lump of blue crystallized sulphate of copper in a porcelain crucible, and warm gently till the water of crystallization has gone. Note change in color. Add a little water. Again note change in color.

Experiment 3.

Change caused by the Action of Sulphuric Acid on Water.

Put about 20^{cc} of cold water in a tt. Add, cautiously, about 5^{cc} of sulphuric acid. Stir with a glass rod. Note the *change in temperature*.

Note. These changes, caused by the removal or addition of water, seem to lie in the borderland between the true physical changes and the true chemical ones. Some chemists think that simple solution [*e.g.*, when salt disappears by dissolving in water, or when sulphuric acid dissolves in water] is always accompanied by a reaction of the substance with the solvent.

Analyses. Syntheses. Metatheses.

Changes of substance, that is, chemical changes, may be divided into three classes, analyses, syntheses, and metatheses.

Changes of Substance by Analyses.

ANALYTICAL CHEMISTRY.

The word **analysis** is derived from the Greek, and means an unloosing. It is applied in chemistry to unloosing the bonds which bind together the constituents of a compound. Those changes in which compound substances are separated into simpler constituents are called *analytical*.

Analysis may be either proximate or ultimate. A **proximate analysis** is one in which a compound is separated into simple constituents, but not necessarily into the simplest; *e.g.*, when we found, by heating, that marble was separated into two oxides — oxide of carbon and oxide of calcium — we made a proximate analysis. Both constituents, however, were capable of further separation. An **ultimate analysis** is one in which the simplest constituents of a compound are determined; *e.g.*, when, by heating red oxide of mercury, we found that it was separated into two substances — mercury and oxygen — neither of which has been separated, we made an ultimate analysis.

Note that already we have made analytical changes in several cases. Review your laboratory work, and now note the analyses made of

SYNTHETICAL CHEMISTRY.

Red oxide of mercury,
Water [hydrogen oxide],
Sulphuretted hydrogen [hydrogen sulphide],
Carbonic dioxide,
Hydrochloric acid,
Marble.

State in every case into what factors each substance was analyzed.

Changes of Substance by Syntheses.

SYNTHETICAL CHEMISTRY.

The word **synthesis** is derived from the Greek, and means putting together. It is applied in chemistry to the formation of compounds. Synthetical chemistry is the opposite of analytical. We have already made syntheses many times. Refer to the proper experiments, and explain the syntheses of

Iron oxide,   Sulphuretted hydrogen,
Water,   Slaked lime,
Sulphurous acid,   Hydrochloric acid,
Sulphuric acid,   Table salt.

State the factors in each case.

Experiment 4.

Synthesis of Chloride of Ammonium.

Make a synthesis of chloride of ammonium from ammonia gas and hydrochloric acid gas. Put a little aqua ammonia in a tt, and a little hydrochloric acid solution in a second tt. Bring the mouths of the two tubes near each other. The gases will join in the air, and white fumes of the chloride of ammonium appear. If the action is not rapid enough, heat each tt to drive the gases from their solutions.

Changes of Substance by Metatheses.

METATHETICAL CHEMISTRY.

The word **metathesis** is derived from the Greek, and means an exchanging. It is applied in chemistry to those changes in which two or more substances change places; *e.g.*, when hydrogen is made by the action of zinc on sulphuric acid, the change is metathetical, for the zinc takes the place of the hydrogen, and the hydrogen is left in a free condition as was the zinc at first. We have already made a great many metatheses. Cite references, and explain the metathetical changes when you made

Hydrogen from sulphuric acid,
Hydrogen from water by means of hot iron,
Zinc sulphate from zinc oxide and sulphuric acid,
Combustible oxide of carbon,
Sulphuretted hydrogen from sulphide of iron and an acid,
Sulphate of magnesium,
Marble powder,
Sulphate of sodium,
Hydrochloric acid from table salt,
Carbonate of potassium from potassium and dioxide of carbon,
Nitric acid.

Experiment 5.

Metatheses.

1. Dissolve about 5^g of nitrate of lead in about 100^{cc} of water. Put the solution in a beaker. Insert a small strip of clean zinc. Note the changing places by the zinc and the lead — the lead lets go its hold of the nitrogen and oxygen part of the nitrate and appears as metallic lead, while the zinc joins that which the lead has left.

2. In a tt dissolve about one tenth of a gram of silver nitrate in about 5^{cc} of water. In a second tt dissolve about one third as much table salt in about 5^{cc} of water. Pour, drop by drop, one solution into the other. Here there is an interchange — the silver leaving its nitrogen and oxygen to join the chlorine which leaves its sodium, while the sodium joins the nitrogen and oxygen left by the silver. The chloride of silver, not being soluble in water, appears as a heavy precipitate.

3. Dissolve about 1^g of potassic sulphocyanate in about 200^{cc} of water in a beaker. Potassic sulphocyanate, on analysis, is found to contain potassium, sulphur, carbon, and nitrogen — the potassium having taken the place of the hydrogen in an acid consisting of hydrogen + sulphur + carbon + nitrogen. Into this solution drop a single drop of ferric chloride. Look for the metathesis as the chloride of iron falls through the solution. What simple substances are there in ferric chloride? Explain the metathesis here.

CHAPTER II.

THE EARLIEST PERIOD.

AN examination of the records of very ancient nations, as the Chinese, Jews, Egyptians, and Phoenicians, shows that these people possessed some knowledge of chemical processes. The Chinese early learned the arts of glass and porcelain making. The Phoenicians are noted for their skill in dyeing. The Jews, as shown by the Old Testament, were acquainted with certainly four, and probably six, metals, some of which could be obtained only from ores, by chemical means. And the Egyptians are famous not only for their knowledge of a large number of chemical processes, but for the skill with which they applied these in their arts. Independently of the Chinese, the Egyptians discovered a method for making glass. It is probable that this discovery was accidental, soda having been added to sand as a flux to aid in separating gold from the sand. It was chiefly from the Phoenicians and Egyptians that the Greeks, and later, the Romans, obtained considerable knowledge of chemical processes.

But all the chemical knowledge of these ancient nations was disjointed, unclassified, and never do we find an attempt at a scientific explanation of chemical phenomena. It is strange that a critical examination of chemical changes could have escaped the keen minds of the Greeks. Their whole attention seems to have been given to the deductive method of reasoning. Seldom did they study Nature inductively,[1] and never do they seem to have tested their deductions by experiments. Though the Ancients never deliberately planned experiments to give an insight into the constitution of bodies, they made numerous speculations as to the nature of the world, and the matter of which it consists. The most celebrated of these is Aristotle's theory of the elements. By elements are here meant the foundation substances of which all the world is made.

Aristotle assumed that there were four of these elements — earth, water, air, and fire. But to Aristotle these words did not convey the same meaning that they do to us. To him they represented different properties that matter itself possesses: earth stood for that which is cold and dry; water for the cold and wet; air, hot and wet; fire, hot and dry. But these four were not entirely sufficient to Aristotle for explaining all phenomena. Hence he added a fifth,[2] called

---

[1] The distinction between the deductive and inductive methods of reasoning should be well fixed in mind. The inductive method is preëminently the method by which the far-reaching developments in all branches of natural science during the last centuries have been attained. The *facts* have first been observed, collected, and arranged. Then from the special cases general principles have been discovered. The deductive method is the method of speculative philosophy. It is the method employed so largely by the Greek and most other philosophers. From general principles deductions are made to fit special cases.

[2] This fifth element of Aristotle became later the *quinta essentia* [whence our word quintessence] of the Alchemists, among whom it caused much trouble. They made many vain endeavors to obtain this, not understanding that Aristotle considered it of the nature of spirit and not matter, and therefore intangible.

aether. This last was all-pervading and of a spiritual nature. Although this theory of the elements is called the Aristotelian it is said to have originated with Empedocles, who flourished about a hundred years before Aristotle,[1] and it is possible that Empedocles himself obtained it from some earlier source, as it is claimed that ancient writings in India declare the world is made of five elements, earth, water, air, fire, and aether; while Buddha considered it made of these five together with a sixth — consciousness.

For Review.[2] Of what nature was the chemical knowledge possessed by the ancient nations? Mention a chemical process known to the Chinese, to the Jews, to the Phoenicians, to the Egyptians. Where did the Greeks and Romans get their knowledge of chemical processes? How did the Ancients' treatment of chemical phenomena differ from our own? What is meant by the deductive method? What by the inductive? Give an illustration of the use of the inductive method. [This illustration may well be taken from "A Chemical Investigation," Part I, of this book.]

[1] Aristotle lived 384–322 B.C.
[2] *To the Student.* After you have read each section of Part III you should at once try to formulate answers to the questions asked. If you cannot make satisfactory answers to all, *re-read the whole section*, looking particularly for answers to the questions that have troubled you. Again formulate answers to all the questions, and, if any are still unanswered, search for the proper answers without re-reading the whole section.

CHAPTER III.

THE PERIOD OF ALCHEMY.

ACCORDING to Aristotle's theory, water is cold and wet, while air is hot and wet. Each, then, has a common property — wetness. It was thought that when the substance water is heated and boils away it becomes air. This was considered a transmutation, that is, a changing of one substance into another. During the years that immediately preceded the Christian era, and during the first years of this era, many men came to think that if a change of the nature just indicated was possible, it must be possible also to change a base metal into a noble, *e.g.*, to transmute lead into gold. It was at this time, it may be said, that chemistry began to have a being. Not that chemistry as a science began to exist so early, but chemistry as a distinct department of knowledge. Up to this time, as has been shown in Chapter II, many nations possessed a knowledge of a number of chemical processes, but it does not appear that any attempt had been made to collect a knowledge of these processes into a distinct class, or to direct a number of them to the attainment of a given end. But when the attention of men became centered on the problem of the transmutation of metals, then it was that chemical processes began to be grouped together and used for the solution of the problem. Even then, however, no attempt was made to group the processes *in any natural series* or to *explain* the phenomena. As late as the eleventh century, chemistry has been defined [in an encyclopedia by Suidas] as "the artificial preparation of silver and gold."

The word *chemeia*, from which comes our word chemistry, has been traced back to the fourth century, but was probably used even before that. Its derivation is uncertain. There is a word *chemi*, an ancient name for Egypt. This word also meant dark, and may have been applied to Egypt on account of the dark color of its soil. The word was also applied to the dark or mysterious portion of the eye, and also, it is said, to a black preparation used in alchemy. Hence it is somewhat doubtful whether *chemeia* meant the Egyptian art, the black or mysterious art, or the art which made use of this preparation. The term alchemy comes from the Greek word *chemeia*, and the Arabic article *al* used as a prefix.

The period of alchemy began with the first attempt at the transmutation of base metals into silver and gold, and extended into the sixteenth century, when chemistry gradually passed into its medical period. No exact date can be set for the origin of alchemy. In searching for its rise, tradition carries us far back among the myths of the past, but historical proof of the practice of alchemy is wanting before the fourth century. It was in Egypt, and toward the close of the fourth century and during the first years of the fifth, that alchemy first attained distinction. In the seventh century, the Arabs overran Egypt and absorbed the chemical knowledge together with many other things possessed by the Egyptians. Early in the eighth century the Arabs advanced and captured Spain. Here they founded universities, and for years fostered learning and the arts. To the Arabian universities in Spain there came many students from the western nations, particularly from France, Italy, and Germany. Here alchemy was studied, and these students, on their return, spread alchemistic ideas among many nations. It may be said that alchemy reached its height in the thirteenth century, and, although in the sixteenth it began to be supplanted by medical chemistry, it did not entirely die out till many years later. In the seventeenth century, Van Helmont says that he changed mercury into gold, claiming to have a substance one part of which could transmute two thousand parts of mercury. In the eighteenth century, too, pieces of metal [usually bronze covered with gilt] were often shown as proofs of alchemistic changes, and even as late as the present century [1] it has been claimed that metals have really been transmuted. In order to see on what kind of observations the alchemists based their hope of transforming metals, let us for the time being put aside all the knowledge of chemical changes which we have gained from our work in the laboratory, and try the following experiments which the old alchemists used to perform.

Experiment 6.

A So-Called Transmutation.

In a small beaker put about 50^{cc} of strong copper sulphate solution. In this immerse a piece of sheet

[1] See Schmieder's History of Alchemy, 1832.

iron.[1] When a deposit has formed on the iron, remove this deposit, press it into a ball, lay it on the desk and rub it with some hard instrument in order to polish it. Note that it is copper. To the alchemists this seemed a change of iron into copper. Copper sulphate solution was obtained, as it may be to-day, from the pools that form in certain mines.

Experiment 7.

Death of a Metal.

Heat in a small Hessian crucible, over a blast-lamp, a small piece of lead. Stir well, to give good air contact, till there is left only a dirty powder. To the alchemists the metal had been destroyed, and the ashes left were the remains from its death.

Experiment 8.

Resurrection of a Metal.

In a small Hessian crucible heat about a gram of oxide of lead, and add several grains of wheat.[2] Stir the wheat, as it chars, into the oxide. Continue heating and adding wheat till globules of molten lead appear. To the alchemists this was the resurrection of the metal. How do *you* explain the transformation?

[1] Nails, stout wire, or other forms of iron will do nearly as well.
[2] It is best first to put the wheat in a covered crucible, or other dish, and heat before use. Otherwise the grains may "pop" and cause annoyance.

Although the transmutation of metals was the most prominent feature of alchemy, there early crept in a second pursuit which soon claimed a large share of attention. This latter was a search for the Philosopher's Stone, a substance of miraculous powers. Not only was it to be the means by which the transmutations themselves were to be effected, but it was to be a cureall for disease, and a bestower of long life and perpetual youth upon its possessor. Many are the claims for the discovery and virtues of this stone — some of them most preposterous. Thus, Roger Bacon claims that it could transform more than a million times its weight of a base metal into gold. Many alchemists who claimed to possess it declared that they had prolonged their lives three hundred, four hundred, and even more years. Even the production of living beings by its means was believed possible.

It is interesting to note that in the period of alchemy we find theories proposed for the composition of substances. Geber,[1] the most famous of the Arabian alchemists, held that there were two elementary substances, mercury and sulphur. As Aristotle's elements do not correspond with our substances of the same names, so Geber's mercury and sulphur must not be mistaken for the substances to which we give these names. Mercury to him was that which produced lustre, malleability, and other metallic properties, while sulphur was that which caused combustibility. Geber believed the metals were compounds, that the noble metals were very rich in mercury while the base ones

[1] He was a physician who flourished in the eighth century.

contained an excess of sulphur. By this theory it did not seem unreasonable to suppose that sulphur might be withdrawn from a base metal, and in this way transformation accomplished. Valentine, the most eminent man of the last years of the period of alchemy, added a third element, salt, to Geber's mercury and sulphur. Salt to him, however, was not what we mean by salt. It was the principle which enabled a body to resist fire and maintain a solid condition. Valentine proposed the use of chemical preparations in medicine, and a little later chemistry and medicine became so much allied that it is customary to speak of the following years as the iatro, or medical, period of chemistry.

For Review. When was the period of alchemy? Give the derivation of the words chemistry and alchemy. What was the original pursuit of the alchemists? What a second pursuit? Where did alchemy probably have its birth? What part did the Arabs play in the development of alchemy? In what way did alchemy spread over the western world? Mention two famous alchemists. What was Geber's theory of the composition of substances? What did Valentine add to Geber's elements? What important step did Valentine propose?

CHAPTER IV.

THE MEDICAL PERIOD.

ALTHOUGH chemical preparations had now and then during the period of alchemy been used in medicine, it was Paracelsus[1] who, during the first half of the sixteenth century, united chemistry and medicine. He boldly maintained that the "object of chemistry is not to make gold, but to prepare medicines." He considered the human body made up of chemical substances, and believed that changes in these substances caused diseases which could be cured by the administration of chemical preparations. Following the lead of Paracelsus, the chief aim of chemists for more than a century was the establishment of medicine upon a chemical basis. The result of this new development in chemistry was of great good to both chemistry and medicine. On the one hand, the properties of chemicals were carefully observed, methods for preparation elaborated, and many new substances found; while, on the other hand, corrosive sublimate, sugar of lead, compounds of antimony, and many other substances previously considered too poisonous to use in medicine, became valuable agents for the physician. It was during this period that the German, Libavius, in 1595, published the first chemical text-book of note — his Alchymia.

Perhaps the most eminent chemist of this period was Van Helmont, of Brussels.[2] He did not accept

---

[1] Paracelsus lived 1493-1541. He was born in Switzerland, but traveled and worked in many lands.

[2] Van Helmont lived 1577-1644.

Aristotle's theory of the elements, nor was he satisfied with that of Geber or of Valentine. He denied that fire had any material existence, and announced that when a metal is treated with an acid and disappears it is not destroyed, proving, as he did by experiment, that a substance continues to exist in its compounds. Of equal importance with his other researches are his observations on gases.[1] Up to this time no distinction had been made between the various gases, such as hydrogen, carbonic dioxide, sulphurous oxide, *etc.*, all being considered air. Van Helmont not only made a distinction between gases and vapors — calling aeriform substances which, when cooled, became liquids, vapors; and those which did not, gases — but also noted the properties of a number of aeriform substances, and distinguished one from another. He studied particularly carbonic dioxide, which he called *gas sylvestre*. This gas he found could be obtained when coal is burned, when beer ferments, by treating limestone or potash with acids, from mineral waters, and in certain caves. That Van Helmont did much to promote the union of chemistry and medicine is shown by the experiments that he carried on with the juices and secretions of the animal body, also by the explanations he gave for the changes which take place within the body. He believed the acid of the gastric juice is the agent which causes digestion, but that if this juice exists in too large a quantity sickness results. To cure this kind of sickness he used as medicine alkaline preparations, while, to cure sickness caused by a lack of gastric juice, he

---

[1] It was Van Helmont who invented the word gas.

administered acid substances. That Van Helmont, though preëminently a medical chemist, still held alchemistic beliefs is shown by an elaborate description he has left of a method for changing mercury into gold and silver. And it may also be said that his work on gases should give him a place in a period which is usually put a little later — the pneumatic period.

During these years of medical chemistry there lived three men who, though they did but little themselves toward bringing about a union between medicine and chemistry, yet deserve to be remembered for their work in practical chemistry — Agricola, a German metallurgist, Palissy, a French potter, and Glauber, a Bavarian chemist.

Agricola was a physician, but while he practiced medicine he found time to study mineralogy and metallurgy in the mines and smelting works of Saxony, many of whose technical products he has described in a book he wrote.

Palissy devoted himself to improving the art of making pottery. He cared little for speculations, and did not believe in the theories of alchemists nor in those of Paracelsus himself. He based his work upon experiments, and although at first he met many disappointments and failures he finally carried his art to a high degree of perfection. Especially well did he succeed in making enamels and in enameling earthen ware, particularly that ware called Faience. The records [1] he has left are characterized by clearness and simplicity.

[1] *E.g.*, L'Art de Terre, in which he speaks of clays, firings, *etc.*, shows how much superior is experiment to theory alone, and gives a most entertaining account of his early struggles and mistakes.

Glauber,[1] who has been called the Paracelsus of the seventeenth century, really devoted much less attention to medical chemistry than to applied chemistry. He enriched pharmaceutical chemistry with many preparations. It is believed that he first obtained hydrochloric acid by treating table salt with sulphuric acid, and first obtained nitric acid by treating nitre with sulphuric acid. It was in the residue from making hydrochloric acid that he found sulphate of sodium, which was called *sal mirabile Glauberi*, and to this day bears the name of Glauber's salt. Glauber was also a writer on economic subjects, frequently urging Germany to make use of its own raw materials and not sell so much of these to other countries only to buy them back when manufactured into various finished products.

In the course of this medical period we see here and there a use made of the inductive method. Experimenting itself was already largely employed, but seldom were theories founded on the results of the experiments. Paracelsus himself spent many of his early years in travel, claiming that the true way for a physician to gain knowledge of real value was not to read books and argue over the precepts of the Ancients, but to examine cases found in his own day and discuss these.[2]

---

[1] Glauber lived 1604-1668.

[2] It is sad to note that this sensible method gained for him nothing but contempt and ridicule from his fellow physicians who clung most tenaciously to the "Authority of the Ancients." At one time Paracelsus was town physician at Basel and here talked so plainly against the impositions practiced by the pharmacists that the latter found means for having him driven from the city, and it is even said that at a later period these same enemies caused his death by throwing him over a precipice.

Palissy, we find, made experiment the sole basis for his work in pottery; while Van Helmont based many of his assumptions on experiment, unfortunately, however, not interpreting his experiments correctly, as, for instance, when he assumed that water was the basis of all organic substances because water appeared whenever he burned these; and, again, when he assumed that only water was necessary for the growth of some plants, because, as it seemed to him, he had been able to make certain plants grow on the surface of pure water.

For Review. Who united chemistry and medicine? When? What advantage came to chemistry from this union? What to medicine? When was the first text-book on chemistry published? What did Van Helmont reject? What deny? What prove? What did he observe in regard to gases? Who invented the word gas? For what is Agricola noted? For what Palissy? For what Glauber? What proof have we that the inductive method was used as early as the medical period?

CHAPTER V.

PERIOD OF ROBERT BOYLE.

To modest, unpretending Robert Boyle, chemistry is so much indebted that we are justified in designating the active years of his life as a distinct period in chemical history. He was born in Ireland in 1627, but spent most of his life in England where he died at London in 1691. It was Boyle who first saw clearly that the inductive method is the only safe method to follow in the pursuit of knowledge. He it was who first gave a proper definition for the term element. And to Boyle is due the establishment of chemistry as a true science. But these three important results are by no means all that came from his labors. The discoveries of Boyle were of particular value in applied chemistry, *e.g.*, his preparation of ruby glass, his discovery of phosphorus,[1] phosphoric acid, *etc.*

Boyle also devoted his attention to the study of gases, as is shown in his work on "The Spring of the Air" and in other publications. His keen observation led him to the discovery of the law [in regard to the effect of pressure on a gas] which bears his name.

---

[1] Phosphorus had already been discovered by Brand of Hamburg, but its preparation was kept a secret and Boyle had to rediscover it.

Experiment 9.

The Law of Boyle.[1]

Take a piece of glass tube 8–10^{mm} bore, of uniform caliber, about 1.5 meter long, closed at one end and bent to form two parallel arms, one of which is at least three times as long as the other. The longer arm must have the open end of the tube. Have ready about 500^g of mercury that is clean and dry.

Note. Dirty or wet mercury will not give a good result, and will render the tube unfit for a second determination.[2] The closest attention to details is necessary in this experiment if a satisfactory result is looked for.

Pour a little mercury into the tube to "seal the bend." Shake the mercury around till it stands as high in one arm as in the other, when the two arms are upright, thus making sure that the confined air is not under any abnormal pressure from an excess of mercury in the long arm. For convenience in adding mercury, it is well to set the tube on the floor. With a piece of string fasten the tube upright to some convenient support, as a knob to a drawer of your desk. From this time on avoid as much as possible any heating of the confined air from contact with the hands or other parts of the body. Why avoid heating? Measure the length of the column of air to be experimented on, *i.e.*, the column

---

[1] See foot-note, page xxvii of the Introduction.

[2] Dirty mercury can often be cleaned by treatment with a few drops of strong nitric acid.

in the short tube. If the tube tapers at all it should be rejected. Why? Allowance should be made if the end of the tube is not closed square across but is rounded, as is usual. Note that there is already a considerable pressure on the confined air, because the whole pressure of the atmosphere [equal to the weight of a column of air directly over the surface of the mercury in the long arm, and extending up as far as the air itself reaches] is exerted on the surface of the mercury in the open arm, and this pressure is transmitted by the mercury around the bend to the lower surface of the confined air. Determine,[1] as follows, the amount of this pressure of the atmosphere: Take a piece of glass tube at least a meter long and 4 or 5^{mm} bore. Heat the tube about 5^{cm} from the end, and draw off a piece in order to leave one end of the long tube closed. Fill the long tube within about two finger-widths of the top with mercury. Put your thumb over the end and slowly invert the tube, letting the big bubble of air pass up, sweeping along the little bubbles. Repeat the inversions till the big bubble has collected all the air it can; then fill the tube completely with mercury, and, without letting any air enter, plunge its open end beneath the surface of mercury held in some stout dish, as a porcelain mortar. Support the tube upright, and note the formation of a "Torricelli's vacuum" at the top. The instrument you have now

[1] If you have a barometer this pressure can be determined readily and more accurately from this than from the crude apparatus made in this experiment. However, no part of this experiment should be omitted, but in the last part it would be better to substitute the reading of a good barometer for the reading made from your own.

made will serve you very well as a barometer, *i.e.*, an instrument for measuring the varying pressure of the atmosphere. The better you have removed the air from the mercury, and the purer your mercury, the more nearly will the readings of your instrument approach those of a high-grade barometer. Measure the height of the column of mercury which the air pressure is able to support. Mercury is 13.6 times as heavy as water. Estimate the pressure of air against one square centimeter of surface. One square centimeter occupies 0.155 square inch. Estimate, then, the pressure *in pounds* against one square inch.

Air pressure is usually expressed simply by the measurement of the length of the column of mercury [in the barometer] which the air supports.

Return to your bent glass tube with its volume of confined air. Add mercury to the long arm till the surface of the mercury in this arm is as many centimeters *above the surface of the mercury in the short arm* as the surface of the mercury in the tube of the barometer stands above the surface of the mercury in the cistern of the barometer, *i.e.*, double the pressure on the confined air in the short arm. Measure the length of the short column now, and note the effect that doubling the pressure has had on the size of the volume. Why would it not have doubled the pressure if you had simply increased the height of the mercury in the long arm by 76^{cm} [more or less] above its own *original height*, and not above the *new height* of the mercury in the other arm?

The Law of Boyle may be stated thus. *The volume of a gas is inversely proportional to the pressure to which it is subjected*, *i.e.*, if the pressure is doubled the volume is halved; three times the original pressure gives one third the original volume, *etc.*

As to the value of the inductive method, Boyle clearly stated that if men really cared to get at the truth there was no way by which they could benefit the world more than by going to work and performing experiments, collecting observations, but not attempting to propose theories till all the phenomena involved had been noticed. From this statement, and Boyle's consistent practice of what it teaches, we see that he was the first to pursue chemistry in a truly scientific spirit. That Boyle wished to establish chemistry as a science independent of medicine, physics or any other, and that up to this time chemistry had not been regarded as a separate science, is shown by his statement that he found most of the disciples of chemistry had hardly any object in view except the preparation of medicines, or the ennobling of metals; that he himself was tempted to enter the art not as a physician or an alchemist, and that with this in view he drew up a scheme of chemical philosophy.

Boyle saw that neither Aristotle's theory of elements nor the theories of the alchemists were sound. He maintained that only *substances that cannot be decomposed* into simpler constituents should be regarded as **elements**; that many of the substances held in his day to be simple would sometime be decomposed; and that one should not attempt to fix any definite number for the elements. A belief like this shows what a long step in advance of all previous chemists the clear-sighted Boyle was able to take. His views in regard to the nature of elements have not been essentially modified to the present day.[1]

In his attempts to separate compounds Boyle did so much analytical work, and devised so many processes for separation and for the recognition of the presence of substances in compounds, that he may be said to have founded the department of chemistry called Qualitative Analysis. It is true that, during the periods of alchemy and medical chemistry, attempts had been made to get at the constitution of bodies, but little advance had been made toward any systematic scheme for separation. It is seldom that separation can be made in so simple a manner as when red oxide of mercury is converted by heat into mercury and oxygen, or as we have done in any of those analyses noted on pages 107–108.

Boyle saw that it is not necessary to separate a metal, a gas, an oxide, from its compound in order to prove that it is present, but was able to tell the presence of substances by certain changes. He found

---

[1] It may be said that at the present date there seems some slight reason to believe that elements chemically similar may sometime be found mutually convertible, and that finally a dream of the alchemists may, in a measure, be realized. Speculation is rife, and speculation based on some semblance of facts, as to the possible separation of *all* so-called elements. That hydrogen may be found to be the basis of all, or that an unknown element of unexpected simplicity may be found a component of all, hydrogen included, has even been suggested by chemists of repute.

that when a solution of a calcium salt is added to a solution containing sulphuric·acid, or a solution of a silver salt added to one containing hydrochloric acid, a white precipitate is caused. By means, then, of calcium and silver salts, he was able to test for sulphuric and hydrochloric acids, respectively; and by means of these two acids, conversely, he tested for calcium and silver salts. He tested for ammonia with the vapors of hydrochloric acid, the production of a white cloud proving its presence. He also made use of plant extracts, as those from litmus, violet, cornflower, *etc.*, in testing for acids and alkalies. He used his plant juices both in solution and on papers as we do still. To this use of "tests" Boyle first applied the name, ever since kept, **Analysis**.[1]

In the practice of Qualitative Analysis at the present day the chemist does not generally isolate substances in order to prove them present, but, like Boyle, uses certain "tests," *i.e.*, he brings about a series of chemical changes, from an inspection of which he is able to judge what substances are present without ever having seen them. During the last hundred years observations have been greatly multiplied and systems of procedure devised so that **Qualitative Analysis** to-day requires, on the part of the analyst, a familiarity with a vast number of changes, a knowledge of when, and in what order these should be brought about, and an ability to interpret their results.

[1] Assaying would be a better term to use for this testing, because this testing is seldom strictly analyzing.

Experiment 10.

Qualitative Tests.

A. Tests used by Boyle.

First prepare re-agents, or test solutions, as follows: [Each solution should be kept in a *clean*, stoppered bottle, or flask.]

I. **Sulphuric acid.** To 20^{cc} of water add 1^{cc} of sulphuric acid.

II. **Hydrochloric acid.** To 15^{cc} of water add 1^{cc} of the strongest laboratory hydrochloric acid solution.

III. **Silver salt.** To 20^{cc} of water add half a gram of nitrate of silver. Shake till the nitrate dissolves.

IV. **Calcium salt.** To 20^{cc} of water add 6 grams of chloride of calcium. Shake till solution takes place. If the solution is not clear, filter into a narrow-necked bottle.

In a tt containing a few cc of I drop a little of IV. Note formation of a white precipitate. Try a similar experiment, mixing some of II with some of III.

B. Tests by Physical Changes.

Take a piece of glass rod, heat one end till it softens, fasten in a piece of platinum wire about 5^{cm} long, and make a loop in the end of the wire. Moisten the loop, and take up a little chloride of potassium. Using the glass rod as a handle, hold the chloride in the flame of the Bunsen burner and note the flame coloration. Remove all chloride from the wire, testing it in the flame to make sure no trace is left; and in the same way note the flame coloration produced by chloride of

CHEMICAL TESTS. 133

lithium, by chloride of sodium, by chloride of barium. The colors here seen are "characteristic," *i.e.*, they indicate to the chemist the presence of the metals, potassium, lithium, sodium, barium, respectively, in the substances tested.[1]

C. Tests by Chemical Changes.

Have ready four tts labeled a, b, c, d. Put a bit of silver in a, a bit of lead in b, a bit of copper in c, a bit of white arsenic in d. In each case the amount of the substance should not be larger than a small pea. Add to every tt about 1^{cc} of nitric acid, and warm till solution takes place. Add about 10^{cc} of water to every tube. Divide every one of the four solutions into two portions, putting part in other tubes labeled a^1, b^1, c^1, d^1, corresponding to a, b, c, d, respectively. To every one of the solutions, a, b, c, d, add a few drops of hydrochloric acid solution. Note effect. Heat the solutions in which hydrochloric acid causes a precipitate. Note effect. Cool. Note effect. To the solutions not precipitated by hydrochloric acid add aqua ammonia till alkaline to test papers. Note effect. Into the solution still unchanged pass sulphide of hydrogen gas. Note effect. Record all observations in the form of a table in the note-book. Note that hydrochloric acid was a characteristic *test* for both silver and lead, giving with silver an insoluble white precipitate, with lead a white precipitate soluble when heated; that, after the

[1] If the laboratory possesses a spectroscope, examine the four flames through this instrument, and map in your note-book the lines seen in each case.

addition of hydrochloric acid had proved there was neither silver nor lead in c, the blue color produced by ammonia was a test for copper; and that, when the absence of silver, lead, and copper had been proved in d, the production of a lemon-colored precipitate, by sulphide of hydrogen, was a test for arsenic.

But for the successful practice of qualitative analysis, the chemist must know not only what tests to use, but in what order the proper tests are to be used. In order to see that this is so, to the solutions numbered a^1, b^1, c^1, d^1, at once pass in sulphide of hydrogen, *i.e.*, use the last re-agent first. Note that here you get three black precipitates about alike in color, solubility, *etc.*, and no distinction is possible.

At the present day there are about seventy substances which resist every effort of the chemist at separation. These are called the elements, but it is by no means certain that some, if not all, may not be separated when we have better means of analysis, or more knowledge to apply the means we do have.

The most commonly occurring elementary substances are the following:

Aluminum,	Gold,	Oxygen,
Antimony,	Hydrogen,	Phosphorus,
Arsenic,	Iodine,	Platinum,
Barium,	Iron,	Potassium,
Bismuth,	Lead,	Silver,
Bromine,	Magnesium,	Sodium,
Calcium,	Manganese,	Sulphur,
Carbon,	Mercury,	Tin,
Chlorine,	Nickel,	Zinc.
Copper,	Nitrogen,	

MIXTURES AND COMPOUNDS. 135

For a complete list of the elements at present recognized, see page 216.

Boyle even proposed a theory that was far in advance of any previous chemical theory. His theory was that matter is made up of small particles — corpuscles; that a chemical compound results from a mutual attraction of the particles when particles of certain kinds of matter come together. Moreover, no one before had stated clearly, as he now did, that a chemical compound is formed by the union of two or more factors, and that the compound itself has properties entirely different from either factor. He distinguished correctly between a mere mixture and a chemical compound.

Experiment 11.

Mechanical Mixture and Chemical Compound.

Not every mixing of substances results in the formation of a true chemical compound.

A. Iron and Sulphur.

Weigh out, of very fine iron filings, 56 decigrams; of flowers of sulphur, 32 decigrams. Grind the two well together in a mortar till the eye can no longer distinguish the separate particles of either substance, and the mass looks homogeneous. Apply a magnet, and note that the iron may be separated from the sulphur. Again grind together, and put the mixture in a large tt. Heat well over a Bunsen flame. Break the tt, and examine the product. Grind in a mortar, and again apply the magnet. Note that no separation can now be made, for one particle seems attracted as much, or as little, as every other. The substance formed is a compound, and one particle has the same properties as every other. What is this compound? What were the factors?

B. **Zinc and Sulphur.**

Caution! In mixing the zinc and the sulphur for this experiment do not grind them in a mortar, as pressure alone is capable of producing the chemical union with explosive suddenness.

With a spatula, mix, as intimately as possible, 65 decigrams of zinc dust and 32 decigrams of flowers of sulphur. Put some of this powder under a microscope and note that combination has not taken place, as the gleaming, dark-colored pieces of zinc may be distinguished from the irregular, somewhat smutted [by the zinc powder] granules of sulphur. Heap the greater part of the mixture on a brick, stone, or some other substance not harmed by heat, and set fire to it. Note phenomena. Examine with a microscope some of the product where the change has been complete. Can you see either zinc or sulphur? What kind of a change has taken place? What caused the change? What did the chemical change itself cause? What did the heat cause?

The simplicity and clearness of Boyle's writings form a pleasing contrast to the self-laudatory works of Paracelsus, the contradictory ones of Van Helmont, and the generally obscure writings of most of the other medical chemists and the alchemists. Robert Boyle was a seeker after the truth. Where he did not himself see clearly he never attempted to deceive by involving his reader in doubt and obscurity. Animated by the spirit of pure investigation, he placed scientific speculation on that sure foundation — experimentation — upon which it has ever since so securely rested.

From the time of Boyle, chemistry has been dependent upon neither medicine, physics, nor any other science, but has of itself been a true science with the solution of a definite problem — the composition of substances — as its goal, and a method — the inductive — both systematic and logical, its means for attaining the wished-for goal. Boyle, too, gave chemistry its highest aim — the pursuit of absolute Truth.

For Review. When did Boyle live? State the law of Boyle. What three important services did he render chemistry? How did he define an element? To what did he apply the term analysis? State one of Boyle's analytical tests. Describe some chemical tests by means of physical changes. Describe a series of tests by means of chemical changes. How many elements are now recognized? Mention the most commonly occurring elementary substances. What theory did Boyle propose to explain chemical changes? Distinguish between a mere mixture and a chemical compound. Upon what foundation did Boyle rest his speculations? What high aim did he give chemistry?

CHAPTER VI.

THE PHLOGISTON PERIOD.

This period, in which the attention of chemists was directed mainly to the explanation of combustion, may be said to cover the eighteenth century. It began, however, even before Boyle's death, in Becher's theory, that when substances were burned, a *terra pinguis*, as he called it, passed off. Stahl,[1] who worked — chiefly at Berlin — during the early part of the eighteenth century, was the chief developer of the phlogiston theory. Stahl believed that all combustible bodies, metals rightly included, contained within them a substance which he called **phlogiston,** and that this substance passed off when combustion took place, and returned when such substances as we call oxides, *e.g.*, red oxide of mercury, black oxide of iron, *etc.*, were, as we say, reduced. Substances which burned vigorously he thought contained much phlogiston, while those which leave little residue after combustion, as coal and soot, he thought to be made of nearly pure phlogiston. According to this theory metals were compounds, while their oxides were of simpler form — one factor, phlogiston, having departed.

It is hard to see how the fact that the residues from burning the metals are heavier than the metals, and consequently, addition, and not subtraction, takes place in combustion, escaped the notice of Stahl. To be sure, his mind was chiefly intent on such cases as the

[1] Stahl lived 1660–1734.

burning of wood, coal, *etc.*, in which the fact that there is great weight to the gases that pass off was then all unknown. A little later attention was called to the fact that residues from the combustion of metals do weigh more than the metals themselves. This difficulty in the way of the phlogiston theory was met by some who assumed that phlogiston was the principle of absolute lightness, *i.e.*, that phlogiston had a negative weight, and so when it was removed from a substance, the substance naturally *gained* in weight; and by others — who simply ignored it. But after the fact that the products of combustion do weigh more than the combustibles had repeatedly thrust itself on the attention of chemists, and after gases had been more carefully studied [in the pneumatic period], it became evident that the theory of phlogiston had fatal weaknesses. Its death blow came in 1783, when, — the true nature of both oxygen and hydrogen gases having been pointed out, also the fact that water consists of these two substances united, — the part that oxygen plays in combustion was proved. To be sure, some chemists — notably Priestly[1] and Cavendish, whose very works did much to destroy the phlogiston theory — vehemently denied, even till the end of the century, and beyond, that this cherished theory was dead.

For Review. State the phlogiston theory. Who developed this theory? When? How was combustion explained by the phlogiston theory? How reduction? What important fact did the phlogistonists either neglect or explain in an absurd manner?

[1] Priestly has been called one of the fathers of modern chemistry, but it has also been said that he was "un père qui ne voulut jamais reconnoître sa fille."

CHAPTER VII.

PNEUMATIC PERIOD.

We may apply the name pneumatic to an important period which immediately preceded the last or modern period of chemistry. The name pneumatic is derived from a Greek word meaning air or wind. It is applied to that period of chemistry in which the attention of chemists was devoted chiefly to a study of gaseous substances. The pneumatic period is over-lapped by that of phlogiston, and even by the medical, for we found that Van Helmont worked with aeriform matter, and, in fact, he is sometimes spoken of as the founder of pneumatic chemistry. He first pointed out the distinction between various kinds of gases, as carbonic dioxide, sulphurous oxide, *etc.*, and it was he who gave us the name " gas " for aeriform matter which does not, on cooling to the ordinary temperature, take the liquid state. Boyle was preëminently noted for his investigation of gases, but neither he nor his contemporaries were able to distinguish between hydrogen and carbonic dioxide, both of which they thought were modified forms of air.

It was Black,[1] who lived one hundred years later than Boyle, who proved the difference between air and carbonic dioxide. He noticed that when carbonates of magnesium or of lime are heated, a gas, identical with Van Helmont's *gas sylvestre*, goes off, and the carbonate

---

[1] Joseph Black, born in 1728, died in 1799, was professor of chemistry at Glasgow and Edinburgh.

loses in weight. He found that what were then called caustic alkalies [our caustic hydroxides], are capable of taking up this gas and becoming what were then called mild alkalies [our carbonates]. Owing to this fixation of the gas he gave it the name of *fixed air*.

Cavendish[1] came soon after Black, and contributed so much to our knowledge of gases that he has been called the father of pneumatic chemistry. To him we owe the pneumatic trough, and he pointed out the necessity for making accurate determinations of the relative weights, or specific gravities, of different gases, in order to distinguish one gas from another.

Let us determine the specific gravities of some gases.

Experiment 12.

Weight and Specific Gravity of Air.[2]

Take a large [2–4 liter] " prescription bottle."[3] Make sure it is perfectly dry and clean. Fit it with a one-hole *rubber* stopper through which passes a short piece of tightly-fitting glass tube carrying a bit of rubber tube fitted with a pinch-cock. Insert the stopper,[4] with its tube, close the pinch-cock, and weigh the bottle, tubes

---

[1] Henry Cavendish, born 1731, died 1810, was an English chemist and physicist, very wealthy and very bashful.

[2] See foot-note, page xxvii of the Introduction.

[3] A cheap, stout, narrow-necked bottle.

[4] In this experiment to make the joints air-tight, it is well to smear the stopper and all rubber connectors with glycerine or vaseline. Be sure, however, that all greasing is done before the first weighing.

and all, *full of air.* With a small air-pump[1] draw from the bottle as much air as you can. Close the rubber tube with the pinch-cock, again weigh the bottle with its fittings, and note the loss in weight. What has caused the loss? Now get the volume of air removed as follows: Have ready a pail, or tub, of water. Open the pinch-cock with the end of the rubber tube well under water. What causes the water to rush in? Lower the bottle in the water till the water inside and outside is on the same level. Why on the same level? Close the pinch-cock, wipe the bottle dry, and weigh on the platform balances [accurately to 1 gram only]. Remembering that 1 gram of water equals 1^{cc}, get the *volume* of the water which has entered, *i.e.*, the volume of the air which was pumped out.

Find the weight of 1^{cc} of air in the room, at the time of doing the experiment. This, then, is the density of air, the **density** of any substance being the weight of a unit volume of that substance. The **specific gravity** of a substance is the number of times heavier a given volume of that substance is than an equal volume of some substance taken as a standard. Water is often taken as the standard substance. If 1^{cc} of water weighs 1 gram, calculate the specific gravity of air referred to water. Calculate the weight of air in a room 15 meters long, 13 meters wide, and 7 meters high. After you get the weight in grams get it in pounds, assuming that 1 pound is equivalent to 453.6 grams.

[1] Small pumps like those often used for inflating the tires of bicycles, serve well for this purpose, *provided they have check-valves for exhaustions.*

THE LAW OF DALTON. 143

Note that air, as well as every other gas, is an elastic body, which can be compressed into smaller space by pressure [see Ex. 9, the Law of Boyle]. The greater the pressure, the greater the amount of air in a given volume, hence the greater the density. If, then, other things being equal, you filled your bottle with air when the barometer stood high, *i.e.*, when, owing to atmospheric conditions, the pressure on the air at the surface of the earth was great, you would get a greater density for the air than you would had the barometer stood low.

The density of a gas is also affected by the temperature. As is well known, the higher the temperature the greater the volume of a given weight of gas, and the lower the temperature the less the volume. The expression of the exact rate at which a gas volume decreases when the temperature is lowered [or increases when it is raised], is called the Law of Dalton, from John Dalton, who seems to have been the first to state this law clearly.[1]

Experiment 13.

The Law of Dalton.[2]

Have ready a 250^{cc} flask fitted with a one-hole *rubber* stopper. Also have ready about 25^{cm} of glass tube to fit the hole in the stopper, also about 25^{cm} of rubber tube to fit the glass tube, also a pinch-cock to close the

[1] This law is also called the Law of Charles, from a Parisian investigator, who is said to have discovered it.

[2] See foot-note, page xxvii of the Introduction.

end of the rubber tube. It is necessary to have ice or snow for this experiment.

Make sure the flask and all tubes are *dry*. Insert the glass tube through the stopper. Let the glass tube reach nearly to the bottom of the flask and protrude only about 1 inch from the stopper. Connect the rubber tube with the glass tube. Let the pinch-cock be placed at the extreme end of the rubber tube farthest from the stopper. Why? Have ready a vessel of warm water. Insert the stopper in the flask. Open the pinch-cock and plunge the flask beneath the water, but do not let any water get into the flask or tubes. Why? Bring the water to the boiling point, — 100° centigrade. When the flask is wholly immersed in boiling water close the pinch-cock and remove the flask from the water bath. Keep your eyes away as the flask may burst. When cool, open the pinch-cock *with the end of the tube under iced water*. When as much water as possible has entered, close the cock. Make an ice-water bath [with *much* floating ice] of your hot-water bath, and return the flask. Have at hand a beaker containing ice-water, in which much melting ice is floating. Again open the cock, with the end of the tube under the surface of the ice-water in the beaker. Keeping the flask well under the cold water, make the level of the water in the flask and that of the water in the beaker the same [by raising or lowering the beaker], and close the cock. Remove the flask from the ice bath. Loosen the stopper, open the cock, and let the water in the tubes run into the flask. Why? Get the volume of the water in the flask.

Also get the total contents of the flask [when the tube is in it]. From the data thus prepared, calculate [a] the volume of the air experimented on at 0°; [b] its volume at 100°; [c] the amount the volume at 0° would expand in going to 100°; [d] the amount it would expand in going 1°; and, finally, [e] the amount that unit volume would expand in going 1°. This last numerical value is called the coefficient of expansion. It should come out about 0.00366 or $\frac{1}{273}$. If 1^{cc} of gas, measured at 0°, loses or gains $\frac{1}{273}$ of its volume for every degree cooled or heated, respectively, at what temperature does it become 2^{cc}? What would be the volume of 1^{cc} of gas, measured at 0°, if cooled to $-273°$? What is the point $-273°$ called?

The Law of Dalton may be stated thus. *The volume of a gas varies directly as the temperature on the* **absolute scale,** *i.e.*, a scale having its 0° point 273° below 0° C., its 273° mark at 0° C., its 373° mark at 100° C., etc. Or the law may be stated thus. *The volume of a gas measured at 0° C. increases [or decreases], by $\frac{1}{273}$ of itself for every degree C. that the temperature increases [or decreases].*

Experiment 14.

Weight and Specific Gravity of Carbonic Dioxide.[1]

Note. In comparing the weights of gases, air, more frequently than water, is taken as a standard substance. The comparison, also, is usually made at what are called standard conditions, *i.e.*, when the temperature

[1] See foot-note, page xxvii of the Introduction.

is 0° C. and the barometer is 760^{mm}. The expression "standard conditions" is often expressed by **N. T. P.**, *i.e.*, Normal Temperature and Pressure. **The weight of 1^{cc} of air at N. T. P. is 0.001293^g.** In determining the specific gravity of a gas, the weight of a known volume is determined, then, by the laws of Boyle and of Dalton, the volume which this weight would occupy under standard conditions calculated, and from these figures the density is determined.

Have ready as large a flask as will ride on your small balance. The flask should be fitted with a one-hole rubber stopper, through the hole of which passes a tightly-fitting piece of glass tube long enough to reach to the bottom of the flask and to project an inch or so from the top of the stopper. The outer end of the glass tube should carry an inch or so of rubber tube with a pinch-cock. All joints should be made air-tight with vaseline or glycerine, but there should be no superfluous grease. Fit the stopper to the flask, fill completely to the pinch-cock with water and measure the contents in cc. Clean and dry the flask, making sure no water lurks in the tubes or at the joints. Loosen the stopper so that air can pass in or out freely, but do not remove the stopper from the flask. Set the flask thus fitted on the pan of the delicate balance. On the other pan set a second flask of the same size closed by a solid stopper. Why? Add any additional weight necessary, to one side or the other, for a perfect equilibrium. Note the temperature close to the flask and take the barometer reading. From the known volume of the flask, calculate, bearing in mind the laws of Boyle and of Dalton, the weight of air which the open flask holds, thus: —

Let $W =$ weight of the air.
  $V =$ volume as measured.
  $t =$ temperature.

Let $H =$ height of barometer.
  $x =$ volume at 760mm [and t].
  $y =$ volume at 760mm and 0°.

$V : x :: 760 : H.$

Therefore $x = \dfrac{V \times H}{760.}$

$x : y :: 273 + t : 273.$

Therefore $y = \dfrac{x \times 273}{273 + t.}$

As the weight of 1cc of air at N. T. P., or 0° and 760mm, is 0.001293g, $W = y$ times 0.001293.

Note. If weights to the amount of the weight of air that fills the flask are placed on the scale with the flask and all the air pumped out of the flask, the flask and its counterpoise will still balance.

Prepare carbonic dioxide from marble and hydrochloric acid. Pass it first through water and then through sulphuric acid to purify it. [Use catch-bottles.] Pass the gas through the rubber and the glass tubes into the flask till a match, held where the stopper is loosened, is extinguished. Insert the stopper, close the pinch-cock, and remove the oxide generator. Open the pinch-cock for an instant to let out any excess of gas. Close it again. Note the temperature close to the flask, take height of barometer, and find the gain in weight of the flask. To make sure the flask is full of carbonic dioxide, again pass in the gas, for five minutes, and proceed as before. If there is any further gain in weight the flask was not full the first time, and you must again pass in the gas for five minutes, and so continue till there is no further gain. This weighing, and passing, and weighing again, is called *weighing to constant weight*. When you have got the weight constant, again take temperature and pressure. The gain in weight, *plus the weight of the air the flask held*, is the weight of the carbonic dioxide at the observed temperature and pressure. From the weight of the carbonic dioxide calculate what would be the weight of the same volume if the temperature were zero and the pressure 760^{mm}, remembering that the *weight* of a given volume of gas varies *directly* as the pressure and *inversely* as the temperature on the absolute scale. The density, then, of carbonic dioxide — *i.e.*, the weight of 1^{cc} — under standard conditions, is found by dividing the weight of the gas, at N. T. P., by the volume of the flask. Then the density of carbonic dioxide at N. T. P. divided by the density of air at N. T. P. [0.001293] gives the specific gravity of carbonic dioxide referred to air.

Experiment 15.

Weight and Specific Gravity of Hydrogen Gas.[1]

In a similar manner to that of the last experiment, determine the specific gravity of hydrogen gas.

Caution! Have no fire near when doing this experiment. Be sure the hydrogen is pure and dry. It is best to pass the hydrogen through at least three catch-bottles, — the first containing a solution of sodium

---

[1] See foot-note, page **xxvii** of the Introduction.

hydroxide, the others sulphuric acid. Pass in a large and rapid stream of hydrogen in order to *fill the flask completely*.

Note that **hydrogen is 14.37 times lighter than air.** As hydrogen is the lightest gas known, it is most often taken as the standard substance for specific gravity determinations in the case of gases and vapors. Calling the sp. gv. of hydrogen unity, calculate [from the data of Experiment 14] the sp. gv. of carbonic dioxide referred to hydrogen.

Experiment 16.

Weight and Specific Gravity of Illuminating Gas.[1]

Take the gas from the gas-burner tap. Do not have an explosion. Proceed as in the previous experiment. Get the sp. gv. referred to water, to air, to hydrogen.

The above method can be used for vapor density determinations, *i.e.*, for determining the specific gravity of substances that are commonly solids or liquids, but which, by heat, can be converted into vapors. For such determinations, of course, the flask must be immersed in a suitable bath [as one of air, paraffine or oil], which can be heated; and instead of filling by passing in vapor, it is better to put a little of the substance in the liquid [or solid] form in the bottom of the flask where the heat of the bath will convert it into vapor, the excess of which must be allowed to

[1] See foot-note, page xxvii of the Introduction.

escape, leaving the flask exactly full of vapor at the moment the tube is sealed and the temperature of the bath and the air pressure determined.

As every gas has its own particular specific gravity, a determination of specific gravity is an excellent method for distinguishing one gas from another, and for telling whether a given body of a known gas contains an admixture of another gas or not.

One of the most important problems presented to the chemists in the pneumatic period was the composition of our atmosphere — was it to be considered a simple substance, a compound, or a mixture of two or more ingredients? The problem was solved, independently, by two of the most eminent workers of the time, Priestly,[1] an English clergyman, and Scheele,[2] a poor Swedish apothecary. The brilliant researches of Priestly added much to chemical knowledge, but, above all, he will ever be remembered for his discovery, in 1774, of oxygen gas. Nitrogen had already, in 1772, been obtained by Rutherford, and immediately after his own discovery of oxygen, Priestly arrived at the right conclusion in regard to the nature of air, *i.e.*, that it is a mixture of two gases.

Priestly failed, however, to find a true explanation for combustion, doubtless, because he clung so tenaciously to the theory of phlogiston. He, too, as well as Cavendish, never, to his dying day, was convinced of the absurdity of a belief in phlogiston.

[1] Joseph Priestly lived 1733–1804.
[2] Karl Wilhelm Scheele lived 1742–1786.

It is interesting to note Priestly's ideas of original investigations, differing, as they do, so much from those now held. It has been said that "he believed that all discoveries are made by chance and he compares the investigation of nature to a hound, wildly running after, and here and there chancing on game [or as James Watt called it, his random haphazarding], whilst we would rather be disposed to compare the man of science to the sportsman, who having, after persistent effort laid out a distinct plan of operations, makes reasonably sure of his quarry."[1]

Scheele, however, formed a striking contrast to Priestly in his methods of investigation, for of Scheele it has been said truly, "his discoveries were not made at haphazard but were the outcome of experiment carefully planned." Though living in a country where he had little scientific companionship and where it was hard to obtain apparatus and materials for his researches, and though he himself was often hard-pressed by poverty, Scheele contributed an immense amount to the rapidly increasing store of chemical knowledge. As examples, from the long list of his works, may be selected: his devising new ways for preparing medical chemicals; his qualitative analyses of many new minerals, from one of which he got molybdic acid; his proof that plumbago is chiefly carbon; his discovery of an acid [now called lactic] in sour milk; his preparation of mucic acid from milk sugar; his discovery of glycerine and many acids in the vegetable kingdom, *e.g.*, tartaric, citric, malic, oxalic, and gallic; his discovery, in an investigation

[1] Roscoe and Schorlemmer, Treatise on Chemistry, vol. I.

of Prussian blue, of hydrocyanic acid, whose properties he described, speaking of its smell and taste, but totally ignorant of its frightfully poisonous nature; his preparation of a beautiful bright green coloring matter, ever since called Scheele's green [1]; and, above all, his extended work on the black oxide of manganese. Scheele died at the early age of forty-four, but the number of discoveries in that short life is, perhaps, unprecedented. *A remarkable power of observation, an extreme diligence, and an ability to plan experiments* which should bear directly on the question in hand and give decisive results in the quickest and simplest way, have given Scheele a high rank among chemists of all lands and all times. He constantly sought the truth by all the means at his command; he was never content to leave anything in doubt which could possibly be proved by experiment; nor was he ever satisfied to let his investigation of any compound rest until he could both take it to pieces and put it together again. To attain his ends, he begrudged no amount of labor on his own part.[2] In his investigation of the black oxide of manganese, then called *magnesia nigra*, he discovered no less than four substances, baryta, manganese, chlorine, and oxygen. It was a strange coincidence that Scheele in Sweden,

---

[1] Prepared by adding sodium arsenite to a copper solution, copper arsenite being thrown down as a grass green precipitate. This substance has been used largely for a paint, but as its poisonous character is now well known it is being superseded by other, *e.g.*, aniline, greens.

[2] It is supposed that Scheele's constant application to his work, particularly at night, brought on a sickness which, aggravated perhaps by actual want, caused his early death.

independently of Priestly in England, should discover such an important substance as oxygen in the very same year, 1774, and also arrive at the same conclusion in regard to the composition of the air. They concluded, as Priestly stated it, that "the air must be made up of elastic fluids of two kinds." Both Scheele and Priestly first prepared oxygen from red oxide of mercury.

But neither Scheele nor Priestly applied his knowledge of the nature of air to an interpretation of the phenomena of combustion, a true explanation of which was reserved for the French physicist and chemist, Lavoisier.[1] Though the discovery was not long delayed, an important intermediate step had to be taken. Cavendish isolated hydrogen and called it "inflammable air." Soon after the discovery of oxygen, he found that a mixture of hydrogen and oxygen when exploded produce water. This was a very important discovery, for hitherto water had been considered a simple substance. As early as 1770, Lavoisier had conducted an elaborate experiment which disproved the theory that by long boiling water could be converted into earth. He boiled water for many days in a closed vessel, having first weighed the vessel and the water. At the end of the boiling he did find a considerable amount of earthy matter in the water, but, as the gain in weight of the water corresponded to the loss in weight of the vessel, he rightly concluded that the water had dissolved some of the earthy matter of the vessel.

[1] Born in 1743, impeached under the Reign of Terror at the time of the French Revolution, Lavoisier was executed in May, 1794.

Lavoisier was a physicist, and as such had an extreme love for the nicety of physical measurements. It was his application of the balance to a study of chemical changes that enabled him to use the discoveries of Priestly, Scheele, Cavendish, and others in arriving at the true explanation of the process of combustion, and thus changing completely the prevailing ideas in regard to the nature of combustion, as well as those about the respiration of animals.

It was in 1770 that Lavoisier published an account of his celebrated experiment on the change of water into earth. Soon after this, he turned his attention to the study of combustion phenomena, particularly those in which a metal on being heated in contact with the air forms a non-metallic substance. For this work he made use of a delicate balance, weighing the substances to be burned and weighing the products of combustion. At first, Lavoisier was a believer in phlogiston, but he soon threw over this belief. Though the phlogistonists eagerly grasped at the newly-discovered hydrogen and attempted to identify this as their long sought for phlogiston, and though it did seem as though hydrogen might be phlogiston, inasmuch as when hydrogen is added to the oxide of a metal, the metal appears again; still there was one important fact which no theory of phlogiston had been able to explain, *i.e.*, the production of quicksilver when the red oxide of mercury is heated apart from hydrogen, charcoal, or any other substance.

As early as 1772, Lavoisier felt so sure that he was on the track of something new in regard to combustion, that he sent to the French Academy a sealed note containing a description of his work up to that time. In this note he stated that when sulphur, phosphorus, and some of the metals are burned, the increase in weight is caused by the absorption of air. A hundred years before this, Mayow[1] had arrived at the conclusion that our atmosphere [and nitre also] contains a substance which is able to unite with metals when they are heated, and is also used by the lungs when we breathe, changing the spent blood into the fresh arterial. This substance, contained in the air and in nitre, he called *spiritus igno-aereus* or *nitro aereus*. Had Mayow lived, it is possible that he, instead of Lavoisier, might have made the great generalization which fell to the lot of the latter.

In his note to the French Academy, Lavoisier stated that when litharge [an oxide of lead] is heated with coal in a closed vessel a great volume of air is produced. It was not long, however, before he himself saw that it was not air, but another gas which was produced in this reduction of litharge. As we have seen, the discovery of oxygen came in 1774. Lavoisier, unhampered by any desire to perpetuate the phlogiston theory, saw at once that the discovery of oxygen gave the key to the explanation of the most important chemical question of the times. He saw that, if you assume that in combustion this gas combines with the metals, charcoal, sulphur, *etc.*, you can explain the increase in weight as well as all other changes without having to resort to any such absurdity as a substance with a

[1] John Mayow, born 1645, died in 1679 at the early age of 34.

negative weight. In 1775 he published his opinions in regard to the action of oxygen; in 1776 he showed that when the diamond is burned oxide of carbon is produced; and in 1777 he proved the amount of oxygen, by volume, in the air by burning phosphorus in a closed vessel, and noting that one fifth of the air had disappeared after the burning.

Briefly stated, Lavoisier's oxidation theory was that substances burn only in *extremely pure* air; that when a substance is burned, for the increase in weight there is a corresponding decrease in the weight of the air; also, that the substance burned is changed into a substance which is generally an acid, though the metals yield substances which are not acid. From this it will be seen that Lavoisier thought the acids contained oxygen. He tried to prove that oxygen itself was an acidifying principle, believing, for instance, that sulphuric acid consisted of sulphur and oxygen, and phosphoric acid of phosphorus and oxygen. Even hydrochloric acid he thought must contain oxygen, and, assuming that when hydrogen is burned an acid must result, he attempted to find the acid. In 1783, however, Cavendish discovered that water is the product of the combustion of hydrogen. Then Lavoisier made use of this discovery of Cavendish's in explaining correctly the manner in which hydrogen is produced from water by red-hot iron; the cause for the appearance of water when oxides are reduced by hydrogen; and the production of hydrogen from acids by adding them to metals.

These important explanations led to such great reforms in chemistry that the admirers of Lavoisier have sometimes claimed that he was the founder of scientific chemistry. That chemistry was a science long before this, however, we have already seen.

To Lavoisier we must give the credit of first discovering a law of general application in chemistry. Never before his time do we find a man thoroughly impressed with the idea that no matter is lost however great the change in which it is involved. He, however, seems to have had this belief constantly fixed in mind, and as constantly to have directed his experiments toward proving that the sum of the weights of the factors, in a chemical change, is exactly equal to the sum of the weights of the products. This great generalization is called **The Law of Conservation of Mass,** or of the **Indestructibility of Matter.**

Experiment 17.

Conservation of Mass.

The Law. *The sum of the weights of the products of a chemical change is exactly equal to the sum of the weights of the factors.*

A. The Combustion Products of a Candle.

Take a piece of fine iron gauze. Cut from it a circular piece about three and one half inches in diameter. Take a square foot of the same gauze.[1] With the square foot make a cylinder around the

---

[1] One piece of gauze will do for a large class of students.

circular piece, so that the circular piece shall form a shelf half way down the cylinder and divide the volume of the cylinder in two equal portions. The cylinder may be kept in place around the disk by being wound with wire.

Place a five-inch piece of paraffine candle on one pan of a platform balance. Set the cylinder over the candle. In the upper compartment put 150–200^g of hydroxide of sodium[1] in stick form. Balance the apparatus with any convenient tare. Light the candle and let its flame pass up among the sticks of hydroxide. At the end of five minutes note the gain in weight. Put out the flame and wait five minutes more. Note that during the second five minutes there is a slight gain in weight, due to the absorption of moisture from the air by the deliquescent hydroxide of sodium. This gain, however, is only a small fraction of the amount gained when the candle was burning. Therefore, though the candle was gradually disappearing, the substance of the candle was not annihilated. What have we already found, by means of lime water, to be one of the products when a candle burns in air? Hold a clean, cold, and dry tt directly over the flame of a candle, but not near enough to the flame to touch it. What is a second product of the burning? Analysis would show that the candle is composed of a substance — paraffine — which is itself composed of carbon and hydrogen only. What happens to hydroxide of sodium

[1] As soon as one student has used this hydroxide of sodium it should be put in a bottle, for, if kept away from the air, it will serve for eight or ten students. The circular shelf of gauze, however, should be washed after every trial.

when in contact with dioxide of carbon, when in contact with water? Why was there an actual *gain* in weight in this experiment?

B. **The Weight of the Products is Equal to the Weight of the Factors.**

Take, of thoroughly dry nitrate of barium, exactly 10^g and, of thoroughly dry sulphate of potassium, exactly 7^g. Dissolve each substance in a separate beaker containing about 100^{cc} of water. Bring both solutions just to a boil. Add one to the other, washing the last drops from the beaker by means of the wash-bottle. Note the metathesis. The barium changes place with the potassium, and there results sulphate of barium and nitrate of potassium. Let the sulphate settle. Have ready a weighed filter paper in a 3-inch funnel. Decant the clear liquid, through the filter, into a good-sized weighed beaker, or evaporating dish. Transfer the precipitate to the filter, washing out every bit by means of the wash-bottle. Do not lose any precipitate, and save all the wash water. Collect all the filtrate, and evaporate, cautiously, to dryness. Get the weight of the residue. Also dry the precipitate and get its weight. Compare the sum of the weights of the two products with the sum of the weights of the two factors.

For Review. What is the meaning of pneumatic? To what period of chemistry is it applied? What did Van Helmont contribute to this period? What did Boyle contribute to our knowledge of gases? What did Black contribute? State the law of Dalton [or Charles]. What is the absolute scale of temperature? When did Cavendish live? What do we owe to him? What is the distinction between density and specific gravity? How may gases be distinguished from one another? Who determined the composition of the atmosphere? When was nitrogen discovered? When was oxygen discovered, and by whom? What is Scheele's green? What enabled Scheele to accomplish so much in so few years? When, and by whom, was the composition of water proved? Who was Lavoisier? What did Lavoisier apply in chemistry? State Lavoisier's theory of combustion. What was Lavoisier's theory in regard to acids? State the first great law of chemistry. What is this law called? Give an illustration of this law.

CHAPTER VIII.

THE MODERN, OR ATOMIC THEORY, PERIOD.

§ 1. John Dalton[1] proposed the most celebrated theory chemistry has ever had — a theory which has been a safe guide for its followers ever since. Soon after Lavoisier established the law of conservation of mass, two other fundamental laws of chemistry were discovered, and it was while pondering over these last two laws that Dalton formed an hypothesis in regard to the constitution of matter which soon developed into his famous Atomic Theory.

During the last decade of the eighteenth century, Richter,[2] a German chemist, investigated the neutralization of acids with alkalies, and stated that when several portions, all equal in weight, of an acid are taken and neutralized, each with a different alkali, the amounts of the alkalies are equivalent. He also believed that the composition of a substance is constant, *e.g.*, that whenever nitre is made by neutralizing nitric acid with hydroxide of potassium the amount of potassium in a given weight of the nitre is always the same, and that when zinc sulphate is formed by the action of zinc on sulphuric acid the percentage of zinc in the resulting sulphate does not vary whether a large or a small amount of acid has been used.

There was an eminent French chemist, however,

---
[1] Lived, 1766-1844, in England.
[2] Jeremias Benjamin Richter, born in 1762, died in 1807, worked in Breslau and at Berlin.

Berthollet,[1] who, in the first decade of the nineteenth century, advanced ideas very much opposed to those of Richter. Berthollet did not believe that compounds possess a fixed composition. He maintained that when a compound is formed from two simple substances, the amount of either constituent in the resulting compound varies according to the quantity of that factor at hand, *e.g.*, that if an acid be treated with a large amount of a metallic hydroxide, the resulting salt will contain more metal than it would had a smaller proportion of the hydroxide been used, and *vice versa*.

On account of the above assumption, Berthollet soon became involved in a dispute, which lasted for eight years, with another chemist, Proust.[2] Proust had already shown that a number of substances never vary in composition. He now undertook to prove the falsity of Berthollet's assumption, and in the end succeeded in showing that his opponent had often analyzed mixtures, instead of true chemical compounds. Very likely Berthollet himself had been led astray by the now well-known fact that some substances form with others a series of compounds, *e.g.*, sulphur, we have found, forms with oxygen sulphurous oxide and sulphuric oxide, carbon also, we saw, forms two oxides, and nitrogen forms not less than five. While Berthollet believed that metals form oxides *with gradually and indefinitely increasing amounts of oxygen*, Proust proved that the formation goes on *in jumps*, *i.e.*, that a given substance, capable of forming two oxides, after it has taken up a given amount of

---

[1] Claude Louis Berthollet, born in 1748, died 1822.
[2] Born 1755, died 1826.

DEFINITE PROPORTIONS BY WEIGHT. 163

oxygen and formed the first oxide, will not, as the conditions are varied, take up a little more and then a little more oxygen till complete saturation is reached, but will jump at once from the first proportion to the last. As the result of this famous controversy a second great law of chemistry became established. This law may be stated thus. *Every distinct chemical compound has a fixed and unalterable composition.* This is called **The Law of Definite Proportions by Weight.**

Experiment 18.

Law of Definite Proportions by Weight.

Weigh out exactly 10 grams of sal soda. The sal soda should be in good clear crystals, not effloresced. Put the sal soda in a beaker tall enough to avoid subsequent spattering out. Dissolve in 40–50^{cc} of water. Add hydrochloric acid solution till no more *effervescence* takes place. Transfer the liquid to a weighed evaporating dish, washing the last traces from the beaker into the dish with pure water from the wash-bottle. Evaporate to dryness, avoiding all spattering. Get the weight of the residue.

Again take exactly 10 grams of sal soda, treat with hydrochloric acid as before, but when the effervescence has ceased add a considerable excess of hydrochloric acid. Evaporate, and weigh. Compare the weight with the first. Has there been, in the second case, **a taking up** of an excess of the acid principle?

§ 2. Quantitative Analysis.

The branch of chemistry called **Quantitative Analysis** is based upon the law of definite proportions by weight. In Qualitative Analysis, the chemist has to determine what substances are present in the body to be analyzed. In Quantitative Analysis, he determines how much of a substance, known to be present, is present. As in Qualitative Analysis, so in Quantitative, the substance looked for can sometimes be isolated, *e.g.*, when we desire to know the amount of mercury in a given weight of the red oxide of mercury, we can heat the substance and weigh the resulting mercury; or if we wish to know the amount of iron in iron oxide we can pass hydrogen over the hot oxide, and after the hydrogen has carried off all the oxygen we can weigh the remaining iron; or when we have passed the electric current through a compound in solution, often one of the constituents is deposited, or driven off as a gas, and this we can collect and weigh. But in a vast number of cases, when quantitative analyses are called for, no such isolation can be made. In these cases certain chemical changes are brought about which convert the whole of the substance looked for into some substance *whose percentage composition is known, or can be found*, and whose weight can be determined accurately. For instance, we desire to know the amount of chlorine in a gram of the salt just made in Ex. 18. It would be very hard to isolate this chlorine from its sodium, and next to impossible to weigh it accurately if isolated. But we can easily combine the chlorine with silver [while

QUANTITATIVE ANALYSIS. 165

at the same time we combine the sodium with nitrogen and oxygen to form nitrate of sodium] and thus convert the chlorine all into silver chloride, which we can weigh. If, now, we determine the percentage of chlorine in silver chloride, as we can do in a simple manner, we can, by arithmetic, calculate the weight of chlorine in the silver chloride that was formed from our original gram of salt, for the law of definite proportions by weight declares that every compound has a fixed and unalterable composition, hence the percentage composition of silver chloride made in one way must be the same as that of silver chloride made in any other way. Knowing the weight of chlorine in the silver chloride made from our salt, we know the weight of chlorine in the salt because all the chlorine of the salt went into this silver chloride.

Experiment 19.

Analysis of Table Salt.

What simple substances have we already proved to be in table salt? How did we prove this?

Have ready about 5^g of c.p. sodium chloride. To make sure that the salt is thoroughly dry, pulverize it, put it in a porcelain evaporating dish and heat for about five minutes over the Bunsen burner flame. When cool, weigh out, on the small balances, exactly 1^g of the salt. Put the salt in a medium-sized beaker, add about 30^{cc} of distilled water, and warm till solution takes place. Then weigh out, and place in a similar beaker, 5.0^g of nitrate of silver. Dissolve the nitrate of silver also in about 30^{cc} of distilled water. Heat both solutions till you just cannot bear your hand on the sides of the beakers, then, using a glass rod to direct the stream, pour, carefully, with constant stirring, the salt solution into the nitrate solution. With a stream of water from the wash-bottle, wash out into the main liquid the last traces of the salt solution. Protect the contents of the beaker from the action of sunlight. Continue to heat, gently, till the precipitate has settled. Have ready, in a funnel, a well dried and weighed filter. Decant the clear liquid, down a rod, through the filter. Transfer all the precipitate to the filter. In order to get out the last particles make use of the glass rod[1] and a stream from the wash-bottle. *Wash the precipitate well* with distilled water, and dry[2] it on the paper. Dry to constant weight, but do not burn the paper. Get the weight of the silver chloride formed.

In order to find the amount of chlorine, we must know what per cent of chlorine there is in silver chloride itself. Determine this as follows:

Weigh out, on the small balances, *exactly* 1^g of pure silver. Place the silver in a small beaker, and add about 2^{cc} of nitric acid diluted with two or three times its volume of water. Heat, gently, to produce rapid solution. When the silver is all changed to the soluble nitrate, add about 25^{cc} of distilled water. In a second beaker dissolve in about 30^{cc} of distilled water, at least 2^g of c.p. chloride of sodium. Heat each solution; mix;

---

[1] For this work it is well to have on the end of the glass rod a piece of tightly fitting rubber tube about 1^{cm} long.

[2] See Appendix M.

let settle; filter; wash, dry, and weigh the precipitate, exactly as in the previous part of the experiment. The silver has taken to itself chlorine from the chloride of sodium. The precipitate consists of silver chloride.

The amount that this chloride of silver weighs in excess of the 1^g of silver represents the amount of chlorine the silver has taken, and from these figures the per cent of chlorine in chloride of silver may be found.

Now calculate the amount of chlorine in the 1^g of chloride of sodium, and also the amount of sodium. Let the final results be stated in parts per 100, *i.e.*, in percentages. A good worker should come within 1% of the true amount of chlorine.[1]

For Review. — §§ 1 and 2. Who proposed the celebrated atomic theory? What was it that Richter noted in the last decade of the eighteenth century? What did Richter believe in regard to the composition of substances? What did Berthollet maintain? What did Proust prove? How, probably, did it happen that Berthollet was led astray? State the second great law of chemistry. What is this law called? State the results of an experiment that illustrate this law. What is the object of Quantitative Analysis? How does Quantitative Analysis depend on the second great law of chemistry? Describe, briefly, an experiment that shows how quantitative determinations are usually made in chemistry.

[1] With the best of balances a good worker should come within 0.1% of the true amount.

§ 3. MULTIPLE PROPORTIONS.

Had Proust, in his examination of those cases in which a given substance forms two or more oxides, thought to start with some definite amount of the one substance and calculate ratios between the varying amounts of the second which unite with the first, he would probably have been the discoverer of the third great law of chemistry. For instance, 10^g of carbon produce either 23.3^g of the combustible oxide or 36.6^g of the non-combustible. Now deducting, in each case, the 10^g of carbon, we find that in the first case there has been a taking on of 13.3^g of oxygen and in the second 26.6^g or just twice as much, *i.e.*, the ratio between the two amounts of oxygen, when the amount of carbon remains fixed, is a very simple one — $1:2$. But it does not seem ever to have occurred to Proust to make such a comparison. Dalton, however, saw that it might be done. It was when examining two gaseous hydrogen compounds of carbon — olefiant gas and marsh gas [1] — that Dalton did this. In 100^g of olefiant gas there are 85.7^g of carbon and 14.3^g of hydrogen, while in 100^g of marsh gas there are 75.0^g of carbon and 25.0^g of hydrogen. Now, calling the amount of hydrogen, in each case, any fixed number, as 10, and making proportions,

$$14.3 : 85.7 :: 10 : x \qquad\qquad 25 : 75 :: 10 : x$$
$$x = 60 \qquad\qquad\qquad x = 30$$
$$30 : 60 :: 1 : x$$
$$x = 2$$

[1] The "fire-damp" of the mines. The oxide of carbon which results from the explosion of this gas mixed with air is called the "choke-damp."

we find that the amount of carbon in the first case is to the amount of carbon in the second case as 2 : 1. Dalton also examined the two oxides of carbon, and found, as we have seen, that the amount of oxygen in the one is to the amount in the other as 1 : 2. For Dalton, this simplicity of numbers seemed to have a deep meaning. In order to find the underlying law, he set out to make an examination of other cases, particularly the formation of various oxides of nitrogen. He soon made the discovery of the law of multiple proportions.

Experiment 20.

Multiple Proportions.

A. The Oxides of Sulphur.

We have studied two oxides of sulphur. In 100 parts, by weight, of one there are 50 parts of sulphur and 50 parts of oxygen. In the second, there are 40 parts of sulphur and 60 parts of oxygen. Calculate the amount of oxygen there is, in each case, if the amount of sulphur, in each case, is called unity. Then calculate the ratio between the amounts of oxygen found. Note that the ratio is 2 : 3 exactly.

B. The Oxides of Nitrogen.

There are five oxides of nitrogen.

In the first the nitrogen = 63.6% and the oxygen = 36.4%
" " second "   "   = 46.6% "   "   " = 53.4%
" " third   "   "   = 36.8% "   "   " = 63.2%
" " fourth "   "   = 30.4% "   "   " = 69.6%
" " fifth   "   "   = 25.9% "   "   " = 74.1%

Find the simple ratio for these compounds.

C. The Chlorides of Iron.

There are two chlorides of iron.

The first has 44.1% iron and 55.9% chlorine
" second " 34.4% " " 65.6% "

Find the simple ratio.

The Law of Multiple Proportions may be stated thus: *When varying quantities of one substance join a fixed amount of some other, the varying amounts of the first bear to each other a ratio expressed in simple numbers, as 1 : 2, 1 : 3, 2 : 3, or the like.*

---

§ 4. Dalton's Atomic Theory.

Dalton did not rest on the discovery of his law of multiple proportions. He was eager for an *explanation* of the law as well as for one of the law of definite proportions by weight. As an explanation of the facts embodied in these two laws, he soon proposed one of the most remarkable hypotheses that any branch of science has ever known — remarkable not only for the manner in which it explained the facts known at its inception, but also for the position which it has ever since held as the very foundation of our modern chemical science. The essence of this hypothesis was this: that every simple substance is made up of minute **atoms**, all alike and *all of the same weight;* that every compound substance is made up of minute particles, all alike, and each a collection of atoms of different simple substances grouped in simple and unalterable numerical ratio and chemically united. The atoms, as the name implies, were considered indivisible.

It has long been a favorite discussion among speculative philosophers whether there can be such a thing as an indivisible particle of matter, some claiming that such a substance is not only conceivable, but likely to exist; while others have said that every body we know is capable of division, that each part from the division is capable of subdivision and again these parts are divisible, and so the division may be carried on forever without arriving at a true atom, that is, a particle absolutely indivisible. Herbert Spencer,[1] a philosopher still living, after discussing this question, has concluded that true atoms are inconceivable. But the chemist of to-day finds it convenient to assume that there are such particles and, what is more, he has some very clear ideas in regard to them.

However, let us not consider the present conception of atoms till we have traced Dalton's work a little farther. Dalton was even bold enough to believe that he could get at the [relative] weights of these atoms, or minute particles, which make up all matter. He thought that this end could be attained by noting the *amount* of one substance which combines with a given weight of another in the formation of a compound. For instance, he made an experiment and concluded that 6.5^g of oxygen united with 1^g of hydrogen,[2] in the formation of water, and, *assuming* that the hydrogen and oxygen united atom and atom, he gave to oxygen

---

[1] "First Principles."

[2] Dalton's determination here was not accurate, as the best modern determinations give almost 8^g of oxygen to 1^g of hydrogen.

the atomic weight of 6.5, calling hydrogen unity. But this assumption, that one atom of oxygen joined one atom of hydrogen, was not based on a knowledge of the fact, and was therefore a mere speculation. We shall return to this point.

Dalton prepared a table giving numbers for the atomic weights of several different atoms. These numbers he frequently altered, and himself recognized the necessity for a most careful determination in the case of every elementary substance, of its "combining number," that is, the number expressing the proportion by weight in which the substance joins other substances. Even to the present day these combining numbers have not been definitely fixed, and experiments are constantly being carried on directed to the accurate determination of the combining number for some one or other of the elementary substances.

§ 5. Combining Number.

Let us determine the combining number for an elementary substance.

Experiment 21.

Determination of the Combining Number for Zinc.

We know that zinc will displace hydrogen in acids, the zinc entering into combination where hydrogen was. If, then, we should take a large weighed piece of zinc and let an acid act on it till just one gram of hydrogen had passed off, the loss in weight of the zinc would show

COMBINING NUMBER FOR ZINC. 173

the amount of the latter which had entered into combination in the hydrogen's place. The number expressing the number of grams of zinc here used would be the combining number for zinc, if that for hydrogen is assumed to be unity. Hydrogen is such a light gas that it would require an experiment on a large scale to obtain a whole gram, therefore the experiment can be made better as follows.

Take a 100^{cc} flask. Fit it with a *good* one-hole cork, or, better, a rubber stopper. Have a delivery tube reaching to the pneumatic trough. Put in the flask 10^{cc} of hydrochloric acid solution, and 20^{cc} of water. Take some freshly cleaned c.p. zinc, and, using the delicate balances, weigh out [accurately to centigrams] 1.30^g. This amount will produce a convenient quantity of hydrogen. Have ready, inverted and full of water, in the pneumatic trough, a 500^{cc} flask for catching the gas. When all is ready drop the zinc in the flask and insert the stopper. Let the action run to completeness. Keep all undue heat away. Why? Note whether any water has been sucked back toward the flask. If any has, make the proper allowance in measuring the gas caught. Note the temperature of the air surrounding the flask; also read the barometer. Make the level of the water inside and outside the flask the same. Place the palm of the hand over the mouth of the flask and quickly invert the flask. From the graduate add water to the brim. Note how many cc of gas the flask had collected. **One cc of hydrogen** *at standard conditions* **weighs 0.00009^g.** Calculate the *weight* of hydrogen displaced. [*Note.*] *If the best possible result is desired,* it is necessary to consider the effect of the moisture that the hydrogen has taken up as it passed through the water of the pneumatic trough. We shall soon learn that two gases can occupy the same vessel at the same time, and that the more gases there are in a given flask the more the pressure, — each gas exerting its own pressure against the confining walls. Although the amount of water vapor that any gas, as hydrogen, takes up on its passage through water, is small and consequently the amount of pressure due to this aqueous vapor is small, still it affects the accuracy of the result of an experiment like this. When the hydrogen was collected it was not subjected to the full pressure of the atmosphere because the water vapor bore part of the pressure.

It has been found by accurate experimenters that the part borne by the water vapor is represented

At 10° by 0.9 cm. of mercury. At 21° by 1.8 cm. of mercury.
    11° " 1.0 " " "   22° " 2.0 " " "
    12° " 1.0 " " "   23° " 2.1 " " "
    13° " 1.1 " " "   24° " 2.2 " " "
    14° " 1.2 " " "   25° " 2.3 " " "
    15° " 1.3 " " "   26° " 2.5 " " "
    16° " 1.3 " " "   27° " 2.7 " " "
    17° " 1.4 " " "   28° " 2.8 " " "
    18° " 1.5 " " "   29° " 3.0 " " "
    19° " 1.6 " " "   30° " 3.2 " " "
    20° " 1.7 " " "   31° " 3.3 " " "

Before reducing your observed volume of hydrogen to volume at standard conditions, calculate the true pressure under which the hydrogen itself was when collected. Taking the weight of the hydrogen and the weight of the zinc, and calling the combining number for hydrogen unity, make a proportion and get the combining number for zinc, *i.e.*, the number which represents the number of grams of zinc that take the place of 1^g of hydrogen.

§ 6. Prout's Hypothesis.

While Dalton's theory was in its infancy an anonymous writer[1] proposed an hypothesis, — that all the combining numbers are whole numbers. He assumed that there was one fundamental substance, — hydrogen; that the other elements themselves were composed of various proportions of hydrogen condensed, and, hence, the weights of all other elements must be simple multiples of that of hydrogen. Though the hypothesis that all the weights of the heavier elements are multiples of those of the lightest is a very pleasing speculation, we can to-day regard it only as such. The most trustworthy determination of the combining weights of hydrogen and oxygen, in the formation of water,[2] does not give the simple ratio $1:8$ but $1:7.93$. [Dalton's was $1:6.5$; later, $1:7$.]

For Review. — §§ 3, 4, 5, and 6. A consideration of what facts led Dalton to the discovery of the third great law of chemistry? State the law of multiple

---

[1] The writer was found afterward to be Prout.

[2] Probably no quantitative determinations in chemistry have been carried on with greater care and ingenuity than those directed to finding the combining numbers for oxygen and hydrogen.

proportions. State the essence of Dalton's atomic hypothesis. What is an hypothesis? What is meant by the term atom? State, briefly, the opinions that have been held in regard to the divisibility of matter. How did Dalton think he could find the relative weights of his atoms? What was the weak point of his method? What is meant by the "combining number"? Describe, briefly, an experiment which illustrates a method for determining a combining number. What was Prout's hypothesis?

§ 7. Molecules.

Dalton's atomic hypothesis explained perfectly the law of definite proportions by weight and the law of multiple proportions. For if the atoms of every simple substance have all the same weight, and if the particles of compound substances are made up of the atoms of different simple substances grouped in unalterable ratio, then it must follow that every particle of every compound will have a fixed and invariable composition; while in the formation of two or more compounds, as oxides, from a fixed quantity of one substance and varying quantities of a second, the varying quantities of the second must all be multiples of the weight of the atom of this second substance.

That every particle of a compound is made up of a group of chemically-joined atoms of different kinds, and that the weight of every particle of a compound is the sum of the weights of the atoms of which it is composed, are very important points. Although the statement that the total weight of every particle of a compound is exactly equal to the sum of the weights of its constituent atoms may seem self-evident to us, it was not so in Dalton's time, for it must be remembered that heat was then considered a material substance. Let us also bear well in mind the distinction here made between the particle of a substance and an atom. The smallest particle into which a substance can be divided by physical means, *i.e.*, by any means except a chemical change which destroys the substance itself, is generally called a **molecule**. Simple substances, as well as compounds, have their molecules. A molecule of a simple substance generally has more than one atom in it, but never contains atoms of more than one kind, whereas a molecule of a compound substance always contains more than one kind of atoms.

To express his ideas more clearly, Dalton adopted a set of **symbols.**

○ represented an atom of oxygen.
⊙ " " " " hydrogen.
● " " " " carbon.
○ ⊙ " a particle [molecule] of water.
● ⊙ " " " " " olefiant gas.
○ ● " " " " " carbonous oxide.
○ ● ○ represented a particle [molecule] of carbonic oxide.

---

§ 8. Relative Weight of the Atoms.

We have already said that Dalton assumed that hydrogen and oxygen, in the formation of water, unite atom and atom, and therefore he concluded that the weight of the atom of oxygen is 6.5 if the weight of the atom of hydrogen is assumed to be 1, and, in general, that the numbers expressing the atomic weights are identical with those for the combining numbers of the elements. There are, however, no grounds for such an assumption. If 8 grams of oxygen [let us take the more recently determined (round) number, rather than Dalton's incorrect 6.5] join one gram of hydrogen, either 8, or some multiple or submultiple of 8, represents the true weight of the atom of oxygen. If the two gases join atom and atom, the weight of the oxygen atom must be eight **microcriths**[1]; but if *two* atoms of hydrogen join one of oxygen to form the molecule of water, then, if the oxygen atom is eight times as heavy as the two of hydrogen, it must be sixteen times as heavy as one. If there are three of hydrogen to one of oxygen, the oxygen atom must weigh twenty-four microcriths, if the hydrogen atom weighs one. And, on the other hand, if there are two atoms of oxygen and only one of hydrogen, as the two of oxygen outweigh one of hydrogen eight to one, then the atomic weight of oxygen is only four.

In Dalton's time there were no means for determining which multiple of the simplest combining number for an element should be taken as its true atomic weight. And up to the present day no method has been discovered which will prove conclusively which multiple of a combining number must be selected. But from time to time laws have been discovered which assist us

---

[1] The unit at present used in speaking of atomic weights is the weight of an atom of hydrogen, and is called a *microcrith*, — the word crith standing for the weight of a liter of hydrogen gas.

much in our inquiry. In 1808 Gay-Lussac,[1] a French chemist, published an account of some work that he and Von Humboldt had been carrying on with gases. Gay-Lussac's attention had been arrested by the fact that, when water is formed from hydrogen and oxygen, exactly two volumes of hydrogen to one of oxygen are always required. Examining a large number of chemical changes in which gases are involved, he discovered another important law of chemistry. He found that a simple numerical ratio holds between the *volumes* of two gases that unite to form a compound, and also between the sum of the volumes of the uncombined gases and the volume of the product [when the product is in the gaseous form]. In Part I we made a number of changes in which gases were involved. Note that one jar of oxygen gas yielded one jar of sulphurous oxide; that one bag of carbon dioxide yielded one bag of carbon monoxide; that one bag of carbon monoxide then yielded one bag of carbon dioxide; and that, in the formation of ammonia, five volumes of hydrogen and two volumes of nitric oxide were required as factors, while the products, steam and ammonia, had they been measured, would have occupied two volumes each. In each of these cases the ratio between corresponding gaseous volumes may be expressed by very simple whole numbers, thus: $1:1$, $1:1$, $1:1$, $5:2$, $2:2$, and $5+2=7:2+2=4$. **The Law of Definite Proportions by Volume** may be stated thus: *In any chemical change the relative volumes of the gaseous factors and products bear to each other a simple numerical ratio.*

[1] Born in 1778; died in 1850.

Experiment 22.

Law of Definite Proportions by Volume.

By means of the electric current decompose water held in any suitable vessel. Have ready two small tts. Mark[1] on one a volume of 1^{cc}, and on the other a volume of 2^{cc}. Fill both tubes with water. Invert the 1^{cc} tube over the pole that is giving off oxygen, and at once invert the other over the hydrogen pole. Note the ratio of the volumes of the two gases evolved in any given time.

---

§ 9. THE MOLECULAR THEORY.

Soon after the discovery of the law in regard to volumes, an Italian physicist proposed an hypothesis which is now the basis of the celebrated **Molecular Theory.** In the year 1811, Avogadro made the suggestion that "equal volumes of all substances, when in the state of gas, and under like conditions, contain the same number of molecules." Three years later the French electrician, Ampère, came to the same conclusion, and this hypothesis is now known either as that of Avogadró or of Ampère. Let us see on what facts this supposition in regard to the structure of gases is based.

Consider for a moment the case of water, which so easily passes from its common state of a liquid to that

---

[1] This may be done by pouring the proper amount of water from a graduate into the tt and marking its height on the outside of the glass.

of a gas, — steam. A cubic inch of water forms about a cubic foot of steam. Now three suppositions can be made in regard to the manner in which the cubic inch of water is able to occupy the cubic foot of space when the water has been converted into steam. First, it may be supposed that the cubic inch of water on being heated simply swells out until it occupies the whole cubic foot as completely in the gaseous state as in the liquid. In the next place, it may be supposed that the water is not itself an absolutely homogeneous mass of matter, but that it is made up of minute particles, each exactly like every other, and that when the change of state takes place these particles do not change their size at all, but are simply driven much farther apart. In the third place, it may be supposed that, as in the second, the water is made up of minute particles, but that when the water is converted into steam these particles themselves swell out till they occupy many times more space than before, and in this way the cubic foot becomes completely occupied. In both the first and the last cases there will be no vacant space whatever in the cubic foot of steam, but every part will have its portion of matter, whereas in the second case, if the particles do not change their original size, there will be vacant spaces, — in fact, the vacant spaces will be vastly larger than those occupied by the particles of matter. Before we discuss the relative probability of these three assumptions, let us make an experiment.

Experiment 23.[1]

Spaces between the Molecules.

Have ready a dry Kjeldahl flask fitted with a two-hole *rubber* stopper. Through one hole pass a piece of glass tube. About a cm of this tube should project inside the flask, and about a cm outside, when the stopper is in position. To the outer end of this tube fix 5 or 6^{cm} of rubber tube carrying two pinch-cocks, one near each end of the rubber tube. Let a piece of glass tube rise from near the bottom of the flask, pass through the second hole in the stopper, and then be bent at a right angle. By means of a bit of small rubber tube connect to this bent glass tube the short arm of another piece of glass tube bent thus: ⌐J. Let the short arm of the latter tube, which is to serve for a pressure gauge, be about 30^{cm} long, and the long arm at least 50^{cm}. Have ready a bath of boiling water, which contains some common salt [to raise the temperature above 100°]. Immerse the Kjeldahl flask in the bath, but take care not to plunge the flask in the hot liquid so suddenly that the glass may crack. Clamp the flask upright, and also fasten the gauge tube firmly. Look at the flask and make sure that it is dry. Pass dry air into the flask. A good way to pass in the air is by means of the blast-lamp bellows. Force a slow stream of air, first through a catch-bottle of sulphuric acid, then through the short rubber tube into the flask, while the excess of air escapes through the gauge tube. After the dry air has been passing into the flask for

[1] See foot-note, page **xxvii** of the Introduction.

SPACES BETWEEN THE MOLECULES. 183

five minutes, close the lower pinch-cock and disconnect the catch-bottle. Pour mercury into the gauge tube till the short arm is about one third filled. Open the pinch-cock [for a moment only] to relieve any undue pressure, and mark the height on the short arm to which the mercury rises. Note that the flask is now exactly full of air [gas No. 1]. Put two or three drops [only] of water in the rubber tube, and close the upper pinch-cock. Then open the lower pinch-cock, and squeeze the water down into the flask. Again close the lower pinch-cock. Note that the water evaporates and forms steam [gas No. 2]. Note the effect on the pressure gauge. Add mercury to the gauge till the mercury in the short arm stands again at the mark, *i.e.*, till the volume occupied by the two gases within the flask is the same as that previously occupied by the air alone. Next introduce two or three drops of alcohol or of ether [or both, one after the other], and note how several gases may occupy *the same vessel at the same time.*

Caution! *In working with alcohol and ether take great pains to avoid fire, as both are readily inflammable.*

Let us now consider our suppositions in regard to the manner in which a cubic inch of water, after passing into the condition of steam, occupies a cubic foot of space. By experiment we have just found that when a vessel is *full* of one gas, another, and still another, can be added. Hence it cannot be that the whole space is occupied by the steam, or by any other gas. It is true that it may be thought that the whole mass of the steam may form an elastic body, and that, when another gas is added, this elastic body of steam contracts, and the other gas, another elastic body, occupies half the space. But if any given portion of the contents of the flask be drawn off and examined, it will be found that this portion contains not one gas alone but some of all that have been added to the flask. In order to see that gases mingle and do not push one another aside, proceed as follows: Take a Kjeldahl flask. Heat its bulb well over a free Bunsen flame. Remove the flask from the flame. Drop in a few small crystals of iodine. Cork tightly. Hold the flask to the light. Shake, and note that the colored iodine vapor mingles with the air.

We have, therefore, proved that our first supposition, namely, that the cubic inch of water simply swells out and occupies completely the whole of the cubic foot, cannot be an expression of the truth. We are then forced to the conclusion that the water is made up of particles — the molecules of Avogadro.

Let us try two other experiments that will help us in deciding whether the particles are large, elastic, and side by side in the gaseous condition or are some distance apart, *i.e.*, whether the particles themselves swell up or keep their original size even when in the state of steam.

Experiment 24.

Irregular Expansion of Liquids.[1]

Have ready three small [2 oz. is a good size] bottles, all of the same size, and each fitted with a good one-

---

[1] See foot-note, page xxvii of the Introduction.

hole cork, or, better, a one-hole rubber stopper. Fit to each stopper a piece of glass tube about 50^{cm} long with its lower end flush with the under side of the stopper. Fill the first bottle with water, the second with alcohol, and the third with ether.

Caution! *Keep all fire away from ether and from alcohol.*

Fill each bottle to the brim, and insert its stopper, cautiously, in such a way that the liquid may be driven part way up the glass tube, and that no air may be left in the bottle. Have at hand a vessel of water large enough to hold all three bottles at the same time, and so placed that it may be heated, forming a **water-bath**. Set the three bottles with their contents side by side in the water-bath. Make a mark on each stem, at the point where the liquid stands. Heat the water-bath, and note the expansion of the three liquids. Note that the ether expands by far the most rapidly. When the ether shows signs of boiling remove it from the bath, and set it away, safe from fire. Note that the alcohol expands the next in amount. When the alcohol shows signs of boiling, remove it, and stop heating the bath. Record the results.

Experiment 25.

Regular Expansion of Gases.[1]

Caution! In order to insure success in this experiment it is necessary to have the gases experimented on perfectly dry, to have all joints carefully greased and

[1] See foot-note, page xxvii of the Introduction.

tight, to have the ice-bath thoroughly cooled to 0° C., and to make sure that each determination is carried on with the apparatus in precisely the same position; that each time the delivery tube dips the same distance under the water in the pneumatic trough; that the level of the water in the measuring cylinder is, each time, kept the same distance above the water in the trough at the time the volume of the displaced gas is read; and, in short, that all conditions are the same for each gas tried.

Have ready a 2 oz. bottle fitted with a two-hole *rubber* stopper. Through one hole of the stopper pass a piece of glass tube which shall reach nearly to the bottom of the bottle, and project above the stopper a cm or two when the stopper is inserted in the bottle. Fit to the upper end of this tube a bit of rubber tube carrying a pinch-cock. Through the second hole insert the end of a delivery tube arranged so that when the bottle is placed in the water-bath the delivery tube will reach to the pneumatic trough. Make sure that the bottle is *perfectly dry*. Carefully grease all rubbers to make all joints air-tight. Have ready, in the water-bath, a mixture of crushed ice and water, or snow and water [but no salt]. Use much ice, as a pasty mass is best. This mixture gives a uniform temperature. What temperature? Insert the stopper with its fittings, open the pinch-cock, and set the bottle in the bath. Make sure that the bath wholly covers the bottle and reaches the stopper. Fill the bottle with dry air in the same way that you filled the flask in Ex. 23. When dry air has passed in for five minutes, close the pinch-cock, and disconnect the catch-bottle. Make sure that the bottle is still well surrounded with a pasty mass of ice and water. Open the pinch-cock [for a moment only] to relieve any undue pressure. Place your graduate, inverted and full of water, over the end of the delivery tube in the pneumatic trough. Heat the water-bath till the water boils. If the blast-lamp flame is used the water will boil in a few minutes. Catch, in the graduate, the air driven out by the expansion of the contents of the bottle while the temperature rises from the freezing point to the boiling. Note the number of cc caught.

Again make an iced water-bath. Now pass *dry* hydrogen down into the bottle till the gas which bubbles up from the pneumatic trough will, when caught in a tt, burn well. Proceed exactly as before, and note the amount of hydrogen driven over by the expansion. How does the expansion of hydrogen compare with that of air?

Again make an iced water-bath, and try the expansion of an equal volume of illuminating gas. Take the gas from the gas tap at your desk. Dry the gas by passing it through sulphuric acid.

Finally, what do you say in regard to the expansion of gases?

Had the vapors of water, of alcohol, and of ether been tried for the same number of degrees [100] of temperature, the result would have been the same, but, of course, the initial temperature must have been higher, *e.g.*, 100° instead of 0°, and the vapors could not be caught above water. Why not? Nor could the dioxide of carbon be tested in this way, for it is enough soluble in water to prevent an accurate determination, if measured above water.

Let us consider the meaning of the results of Exs. 24 and 25. We find that every substance in the liquid form has its own special rate of expansion. But if the temperature is so high that the liquids have been changed to gases, the expansion of the substances in the gaseous state is for each about the same. Now it must be the effect of the particles acting each on the other that causes the different rates of expansion for liquids, but, if we assume that these same particles, without change of size, are driven so far apart, in the conversion of the liquids into gases, *that they have little or no effect upon each other*, it seems natural that heat should have the same effect on one gas that it does on another, *i.e.*, that all gases should have the same rate of expansion, particularly, if we assume, with Avogadro, that there are, in equal volumes, equal numbers of the particles. From a consideration of these facts and many others, physicists have come to the conclusion that matter is made up of minute particles, — that, in the state of gas, these particles are not closely packed, but, in fact, that the spaces between them are very large in comparison with the size of the particles themselves; that, in the liquid state, the particles are near together, and influence one another; and that, in the solid state, the particles are still nearer together, and cannot move from their relative positions.

It is, moreover, believed that all molecules are in

constant motion, — that, in the case of gases, they are moving about among themselves, with great velocity; that, in the case of liquids, the motion is much more restricted; and that, in the case of solids, the motion is one of vibration or of rotation, and not a changing of place.

There seems no hope that the best of microscopes will ever be able to show the molecules. The physicists estimate that the diameters of the molecules of some substances, — as alum and albumen, — that have remarkably large molecules, range from the one 10,776,000th of an inch to the one 5,000,000th of an inch. But "the best microscopes made to-day will enable one to see as barely visible a point the one hundred-thousandth of an inch, so that such a microscope would need to be as much more powerful than it now is as one hundred thousand is contained in five millions, that is, fifty times, in order to see the albumen molecule, and for the alum molecule as many times as one hundred thousand is contained in ten million seven hundred thousand, that is, one hundred and seven times. Now, one who is familiar with the microscope would probably admit that one might be made through improved methods of making and working glass hereafter to be discovered, two or three, or even ten times better than the best we have now; but the idea of one being made fifty or a hundred times more powerful than we have to-day, I do not think would be allowed to have any degree of probability. The powers of the microscope have not been doubled within the last fifty years, and I suppose more time and ingenuity have been given to the problem of improving it than will ever be given to it in the same interval again." [1]

To give some idea of the actual size of the molecules of water, it may be added that it has been estimated that if a drop of water should be magnified to the size of the earth, the molecules would appear not larger than cricket-balls, and not smaller than small lead shot.

The great molecular theory, which has sprung from the hypothesis made by Avogadro, serves to explain phenomena, and by its application undiscovered facts have been predicted and verified time and again. Still we must remember that it is only a theory, that we never have seen the molecules, and, after all, there may not be such things as molecules, but this theory is based on facts of a very substantial nature, and we cannot help thinking that in it lie the elements, at least, of some absolute truth. We must, however, in all our work, bear in mind the distinction there is, and always must be, between facts and the hypotheses and theories that are put forth to explain the facts. "When, however," to quote from Professor Cooke,[2] "we come to study the history of science, the distinction between fact and theory obtrudes itself at once upon our attention. We see that, while the prominent facts of science have remained the same, its history has been marked by very frequent revolutions in its theories or systems. The courses of the planets

---

[1] A. E. Dolbear. "Matter, Ether and Motion."
[2] "The New Chemistry." Let every student who wishes to know more about the molecular theory get this book, and study the first chapters.

have not changed since they were watched by the Chaldean astronomers, three thousand years ago; but how differently have their motions been explained — first by Hipparchus and Ptolemy, then by Copernicus and Kepler, and lastly by Newton and Laplace! — and, however great our faith in the law of universal gravitation, it is difficult to believe that even this grand generalization is the final result of astronomical science.

"Let me not, however, be understood to imply a belief that man cannot attain to any absolute scientific truth; for I believe that he can, and I feel that every great generalization brings him a step nearer to the promised goal. Moreover, I sympathize with that beautiful idea of Oersted, which he expressed in the now familiar phrase, 'The laws of Nature are the thoughts of God' . . . Through the great revolutions which have taken place in the forms of thought, the elements of truth in the successive systems have been preserved, while the error has been constantly eliminated; and so, as I believe, it always will be, until the last generalization of all brings us into the presence of that law which is indeed the thought of God."

For Review. — §§ 7, 8, and 9. Show how Dalton's atomic hypothesis explained two of the fundamental laws of chemistry. What is meant by the term molecule? How does a molecule usually differ from an atom? Do simple substances have molecules? Give some of Dalton's symbols, and tell what he wished each to represent. State why Dalton's value for the relative weight of the oxygen atom may have been far from correct. Describe, briefly, the work of Gay-Lussac, which led to the discovery of a fourth great law of chemistry. State the law of definite proportions by volume. Who was Avogadro? When did he make his famous suggestion? Who, soon after, made the same suggestion? What was this famous suggestion? What three suppositions can be made in regard to the change of volume that takes place when a cubic inch of water passes into steam? Show why two of these suppositions cannot be expressions of the truth. Give reasons for thinking that the third supposition is an expression of the truth. What is believed in regard to the movements of the molecules? What is believed in regard to the actual sizes of the molecules?

§ 10. Determining Atomic Weights.

The law of Gay-Lussac, together with the hypothesis of Avogadro, proved of great assistance in seeking to establish the true weights of atoms and of molecules. For instance, we can now prove that the molecule of hydrogen gas contains two atoms of hydrogen. If one volume of hydrogen gas is mixed with one volume of chlorine gas and exploded, it is found that the resulting hydrochloric acid gas occupies just as much space as the two factors before the explosion, *i.e.*, one volume of hydrogen plus one volume of chlorine gives two volumes of hydrochloric acid. This may be represented thus:

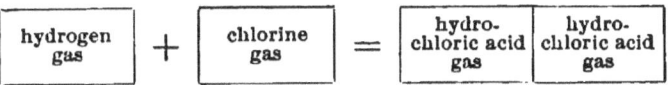

Let us assume that our unit volume here contained just 1,000,000 molecules. Then, by the hypothesis of Avogadro, if there are 1,000000 molecules in the one volume of hydrogen, there must be 1,000000 molecules of chlorine in the one volume of chlorine, and 2,000000 molecules of hydrochloric acid in the resulting two volumes of hydrochloric acid gas. Analysis shows that every molecule of hydrochloric acid contains both hydrogen and chlorine, therefore, in the 2,000000 molecules of hydrochloric acid there must be at least 2,000000 atoms of hydrogen, but these 2,000000 atoms of hydrogen came from only 1,000000 molecules of hydrogen gas, hence we are forced to believe that every molecule of hydrogen has two atoms of hydrogen in it.

If, then, a single atom of hydrogen weighs one microcrith, and we have proved that there are at least two atoms of hydrogen in the molecule of hydrogen gas, the molecular weight, *i.e.*, **the weight of a molecule of hydrogen, must be** at least **two microcriths.** As no proof has ever been presented to show that there are more than two atoms in the molecule of hydrogen, the accepted molecular weight of hydrogen is two microcriths.

Experiment shows, moreover, that two volumes of hydrogen and one volume of oxygen join and form two volumes of steam. This may be represented thus :

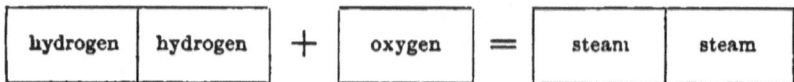

Let us assume, as before, that each single volume contains 1,000000 molecules. Then, as the three volumes are condensed to two volumes, the 3,000000 molecules must, by the hypothesis of Avogadro, be condensed to 2,000000 molecules. Analysis shows that each molecule of steam has in it at least one atom of oxygen. As there are 2,000000 of the steam molecules, there must be 2,000000 oxygen atoms. These 2,000000 oxygen atoms came from 1,000000 oxygen molecules. Hence each molecule of oxygen must have at least two atoms of oxygen. Finally, as there were 2,000000 molecules of hydrogen used, and each molecule, as proved, on page 193, had two atoms of hydrogen, 4,000000 atoms of hydrogen must have gone into the 2,000000 molecules of steam. Hence each molecule of steam must have two atoms of hydrogen. And as there is at least one oxygen atom in each molecule of steam, we conclude that the composition of water [steam] is two atoms of hydrogen and one atom of oxygen. If, then, the oxygen outweighs the hydrogen 8:1, the single atom of oxygen, being eight times as heavy as two atoms of hydrogen, must be sixteen times as heavy as one atom of hydrogen. And **sixteen is the atomic weight we assign to oxygen.** If it is true that there are two and only two atoms of oxygen in the molecule of oxygen gas, what is the molecular weight of oxygen? What is the molecular weight of water?

§ 11. Determining Molecular Weights.

If we accept the hypothesis of Avogadro as an expression of the truth, we can calculate the molecular weights of all substances which are naturally in the gaseous state or can be converted into this state, *e.g.*, water can have its molecular weight taken when it is in the form of steam, and some solid substances, as paraffines, can be heated till vaporized in order to have their molecular weights taken. The determination of molecular weights by means of Avogadro's hypothesis is called the **Physical Method for Molecular Weight Determinations.** It is a very simple method.

If equal volumes of all aeriform substances contain the same number of molecules, the weight of any one molecule must bear the same ratio to that of any other molecule that the weight of a given volume of the first gas bears to the weight of the same volume of the second gas. Having proved that a molecule of hydrogen weighs 2^{mc} [two microcriths], if we get the weight in grams of a given volume of hydrogen gas and the weight in grams of *the same volume* of some other gas, and reckon how many times heavier the given volume of the latter is than the same volume of hydrogen, or, in other words, if we determine the specific gravity of the second gas in reference to hydrogen [instead of to air or to water] as a standard, and multiply by 2, we shall have the molecular weight of the second gas. The reason it is necessary to multiply by 2 is because, although each molecule of the second gas is just as many times as heavy as a *molecule* of hydrogen, as is the whole volume of the second gas than an equal volume of hydrogen gas, yet the molecule of hydrogen itself weighs 2^{mc}, being composed of two *atoms*, each of which contains unit quantity, that is 1^{mc}, of hydrogen.

Let us determine some molecular weights by the physical method.

Experiment 26.

Determination of Molecular Weights by the Physical Method.

A. Molecular Weight of Carbonic Dioxide.

We have already got the sp. gv. of carbonic dioxide referred to air and to hydrogen. See Ex. 14, page 145, and Ex. 15, page 148. Now get the molecular weight of carbonic dioxide.

As coal gas and air, two other substances whose sp. gv. we have determined, are mixtures and not chemical compounds, they have no such things as molecular weights, for they contain molecules of different substances.

B. Molecular Weight of Oxygen Gas.

Determine the sp. gv. of oxygen gas by the method of Ex. 14, page 145. From this result get the molecular weight of oxygen gas.

Although the physical method for determining molecular weights is a very simple and useful one, it is limited in its application to substances which are gases or can be converted into vapors. The molecular weights of other substances are determined by what is called the chemical method. In using the chemical method it is always necessary to get a start by the physical method, *i.e.*, the determination of the molecular weight of some gas is first made from its sp. gv. referred to hydrogen. Then a chemical change is brought about, in which both the substance whose molecular weight it is desired to find and the gas whose molecular weight is known are involved.

If one molecule of the substance whose molecular weight is known always produced one molecule of the substance whose molecular weight is required, or *vice versa*, the application of the chemical method would be extremely simple, for the gram weight of the first would be to the gram weight of the second as is the molecular weight of the first to the molecular weight of the second. But as one molecule of the known often furnishes either more than enough atoms to form just one molecule of the unknown or less than enough [and *vice versa*] the application is difficult and usually requires a study of many chemical changes into which the unknown enters before the number of atoms in the molecule and the correct molecular weight can be fixed.

Experiment 27.

Determination of Molecular Weights by the Chemical Method.

We have already determined, by the physical method, the molecular weight of oxygen. Let us assume that we have found this weight to be 32^{mc}. Now if we bring about some chemical change in which oxygen is involved, we can get the molecular weight of some other substances.

A. Molecular Weight of Chlorate of Potassium.

If your chlorate of potassium is not known to be pure, dissolve 10–20 g. of it in hot water and recrystallize. Powder the crystals and dry at a temperature between 100° and 200° C.

In a weighed porcelain crucible put about two grams of pure dry chlorate of potassium. Get the exact weight of the chlorate. Put the cover on the crucible. This cover should be weighed also, that its weight may be added in case there is any spattering. Heat the chlorate, gently, with a Bunsen burner, but *take care that the contents of the crucible do not foam up and pour over the sides.* After the chlorate has melted continue heating till the mass again becomes solid. Then apply the blast-lamp, gently, till the mass again melts. Just as soon as the second melting has taken place, remove the flame, let cool, and weigh. Again just melt the substance, let cool, and weigh. Continue till the weight is constant. Note the loss in weight caused by the escape of the oxygen.

We can now calculate the molecular weight of chlorate of potassium. In making our calculations, however, we meet with one difficulty. We do not know how many atoms of oxygen each molecule of chlorate has furnished for making the molecules of the oxygen gas. If one molecule of the chlorate has lost one atom, only, of oxygen, the ratio of the weight of the chlorate to the weight of the oxygen gas gone off is x : half 32, or 16, because we have proved that there are *two atoms* of oxygen in the *molecule* of oxygen gas, also that its total molecular weight is 32. If every molecule of the chlorate has lost two atoms of oxygen, the ratio is x : 32, because the two atoms of oxygen are just enough to form a single molecule of oxygen gas. If, however, every molecule of chlorate has lost three atoms of oxygen, the ratio is x : one and a half times 32, or $32 + 16 = 48$. A study of the chemical changes into which chlorate of potassium enters shows that it has three atoms of oxygen, and all are given up in this experiment. Hence in making our calculations we must consider that one molecule of the chlorate has furnished enough oxygen to form a molecule and a half of oxygen gas. Therefore the molecular weight of one molecule and a half of oxygen $[32 + 16 = 48]$ is to the molecular weight of one molecule of the chlorate $[x]$, as is the weight in grams of the oxygen given off, to the weight in grams of the chlorate taken. Make the proper proportion, and get the molecular weight of chlorate of potassium.

B. Molecular Weight of Chloride of Potassium.

From the data of *A* get the molecular weight for the chloride of potassium which was left in the crucible after heating.

Note. It is obvious that if we can find any chemical changes in which either the chlorate of potassium or the chloride enters, and if we can weigh the factors and the products, we can make proportions and calculate the molecular weights for the other substances involved in the changes. We are, of course, sometimes limited, because we cannot always tell just how many atoms leave one molecule, or how many go to some other. But it is seldom that we are left with nothing to help us in our choice between several multiples.

C. Molecular Weight of Sulphate of Potassium.

In getting the molecular weight of sulphate of potassium we shall take advantage of the fact that we have just determined the molecular weight of chloride of potassium, and of the fact that the sulphate can be made from the chloride by the action of sulphuric acid.

Put in a weighed porcelain crucible a small amount [from 1^g to 1.5^g] of finely powdered, dry, c.p. chloride of potassium. Get the exact weight of the chloride used. Add a little sulphuric acid [c.p. is best], drop by drop, while you pass the Bunsen burner flame, now and then, below the crucible. Heat, gently at first, with the Bunsen burner as long as fumes come readily, but avoid all spattering out. Finally, apply the blast-lamp flame till the molten mass becomes a spongy solid. Cool and weigh. Heat to constant weight. As it takes two molecules of the chloride to make one of the sulphate, it is necessary, in making our proportion in this case, to take double the molecular weight of the chloride to compare with the molecular weight of the resulting sulphate.

Find, also, the molecular weight of sulphuric acid, assuming that it takes one molecule of the acid to produce one molecule of the sulphate from the two molecules of the chloride, and that the amount of the sulphuric acid needed to act on 1^g of chloride was 0.65^g. Finally, get the molecular weight of hydrochloric acid, assuming that during the change of the chloride into the sulphate there escaped two molecules of hydrochloric acid gas, and that the weight of this gas produced by 1^g of the chloride was 0.49^g.

Knowing the molecular weights of sulphuric and hydrochloric acids, we could, in a similar manner, get molecular weights for substances with which these acids react.

Neither the physical nor the chemical method for determining molecular weights can be applied universally. The physical method is confined to substances which are naturally gases, or can be converted into vapors at a comparatively low temperature; while the chemical is limited, inasmuch as a start has to be made from some determination obtained by the physical.

For Review. — §§ 10 and 11. Prove [using the law of Gay-Lussac and the hypothesis of Avogadro] that the molecule of hydrogen gas contains two atoms of hydrogen; that a molecule of water [steam] has two atoms of hydrogen; that the weight of an atom of oxygen is 16 microcriths. What is a microcrith? What is a crith? What is the weight [in microcriths] of a molecule of hydrogen? Of a molecule of oxygen? Of a molecule of water? Define the physical method for determining molecular weights. Describe an experiment that illustrates this method. Define the chemical method for determining molecular weights. Describe an experiment that illustrates this method. In what respect is the physical method limited? In what respect is the chemical method limited?

---

§ 12. Specific Heat.

Another aid in determining which multiple of the combining number should be taken for the atomic weight, was found in the relation discovered to exist between the specific heats of elementary substances and their atomic weights.

In the year 1819 this discovery was announced by two French chemists, Dulong and Petit.

The **specific heat** of any substance may be defined as the amount of heat required to raise the temperature of one gram of that substance one degree, *compared with* the amount of heat required to raise the temperature of the same amount [one gram] of a standard substance [water] the same distance [one degree].

Be sure you see clearly the distinction between temperature and heat itself. The temperature of a body is simply its state in respect to heat or cold. What the heat itself is we do not know. The most reasonable explanation of heat is that it is motion of some kind. If, for instance, a piece of iron, as a nail, is pounded vigorously with a hammer, it soon becomes too hot to be held.

Experiment 28.

Transference of Motion.

Take a nail, and having placed it on some hard substance that will not be harmed, pound it vigorously with a hammer till it becomes too hot to be held.

It is believed that at the moment the motion of the hammer is arrested, the particles of the iron receive a kind of motion,[1] which is manifested, as we say, by an increase of temperature. It is not thought that the particles of a solid move around among one another, but that the motion is a vibratory or a rotary one; the greater the vibration or the rotation, the greater the heat. The particles of gases, however, are supposed to change places very rapidly.

It might be thought that the amount of motion, *i.e.*, the amount of **heat**, imparted to a gram of one substance in order to raise its temperature one degree, would, if imparted to a gram of any other substance, cause a rise of just the same extent, but experiment shows that this is not so. When exposed to the same source of heat, some substances reach a given temperature quicker than others. Of all substances, water is the slowest to reach the given temperature; *e.g.*, the amount of heat that will raise a given number of grams of water from 0° C. to 10° C. would raise the same amount of iron from 0° C. to 88° C.; the same amount of mercury to 300° C.; the same amount of silver to 175° C.; and so on.

[1] It is supposed that, even before struck with the hammer, the particles had a considerable amount of this same kind of motion.

The amount of heat that is required to raise one gram of water one degree is called a **calorie**,[1] and is the unit taken for determinations of *quantities* of heat, just as the degree is the unit taken for temperatures. The specific heat of a substance may also be defined as the number of calories required to raise a given weight of the substance a given number of degrees compared with the number of calories required to raise *the same* weight of water the same number of degrees. If the weight selected is a gram for both the water and the other substance, and the distance the temperature is to be raised is one Centigrade degree, then the number expressing the specific heat of the other substance is always a fraction, and less than 1; for it takes but one unit of heat to raise one gram of water 1° C., and it takes less heat for a gram of every other substance.

Let us determine the specific heats of a few substances.

First, prepare a **calorimeter**, *i.e.*, a piece of apparatus for measuring heat quantities. Take a beaker that holds about 200^{cc}, and a second beaker somewhat larger. The second beaker should be of such a size that when the 200^{cc} one is set in it there will be a space of 1–2^{cm} between the walls. Place a layer of cotton-wool on the bottom of the larger beaker; on this cotton set the smaller beaker, and pack *loosely* the space between the walls with cotton. This packing will serve to prevent a considerable loss of heat in subsequent work.

---

[1] This small amount of heat is sometimes called the millicalorie, in distinction from the large calorie more frequently used, which is the amount of heat it takes to raise one kilogram of water one degree.

SPECIFIC HEAT OF ZINC. 205

Experiment 29.

Specific Heat of Zinc.[1]

Have ready the calorimeter just made, a thermometer, and a long, wide tt.[2] Also have ready some water, boiling, in a large beaker, iron pot, or any other suitable vessel. Twist around the neck of the large tt a piece of wire stiff enough to serve as a handle. Pour exactly 100^{cc} of cold water[3] into the inner beaker of the calorimeter, and exactly 50^g of zinc [dust or fine granular] into the large tt. Warm the thermometer somewhat in the steam of the boiling water, and then take the temperature of the boiling water. Cool the thermometer somewhat, and set it in the water in the calorimeter. Heat the zinc to the temperature of the boiling water by plunging the large tt and its charge well down into the boiling water, and keeping them there at least five minutes. Be sure the water continues to boil well. A bit of cotton should be stuffed in the mouth of the tt to prevent spray going in.

With the thermometer stir the water in the calorimeter, and note its temperature accurately to a tenth of a degree. Remove the thermometer; also the plug of cotton from the mouth of the tt. Then at once pour the zinc into the cool water, again insert the thermometer, stir as long as the temperature continues to rise,

[1] See foot-note, page xxvii of the Introduction.

[2] This tt should be about 15^{cm} long, and, if possible, 2^{cm} wide.

[3] When ice or snow or even very cold water is at hand, it is well to cool this water [before measuring out the 100^{cc}] a few degrees below the temperature of the laboratory. Why?

and note the total rise of the temperature, — accurately to a tenth of a degree.

In spite of the cotton packing there will be some loss of heat, and the temperature will not rise quite as far as it should. If at the start the water in the calorimeter was cooled about as much below the temperature of the room as it is heated above that temperature at the end of the experiment, it is fair to suppose that the cold water at the start gains, abnormally, about as much heat as the warm water at the finish loses, and that the result of an experiment performed in this way is more accurate than the result of one in which no such precaution is taken. Then, too, some of the heat from the zinc is lost in heating the thermometer, and some in heating the beaker itself. Owing to such errors as these, the result of this experiment will not compare very well with results obtained where the utmost precautions are taken, and allowance is made for the heat absorbed by thermometer and calorimeter. However, it is easy, with this apparatus, to get results near enough to the truth to form good data for subsequent work.

Knowing the amount of water that was heated, and the amount of zinc that gave the heat, and having noted the number of degrees that the temperature of the water rose, and the number that the temperature of the zinc fell, we can calculate the specific heat of zinc. Bear in mind that a calorie is the amount of heat it takes to raise one gram of water one degree, and state:

[1] The number of units of heat that the whole of the water gained.

SPECIFIC HEATS. 207

[2] The number of units of heat that the 50^g of zinc lost.

[3] The number of units of heat that the 50^g would have lost if they had fallen only one degree of temperature.

[4] The number of units of heat that *one* gram of zinc would have lost in falling *one* degree.

As it would take just as much heat to raise one gram of zinc one degree as it lost in falling one degree, the number obtained for [4] must represent the specific heat of zinc.

Experiment 30.

Specific Heat of Iron.[1]

In a manner similar to that of Experiment 29, get the specific heat of iron. Use iron in the form of filings or small nails.

Experiment 31.

Specific Heat.[1]

Get the specific heat of either copper [use bits of wire]; lead [use 100^g of shot]; or mercury [use 100^g of mercury, and have only 50^g of water in the calorimeter].

The discovery made by Dulong and Petit was that, in the case of simple substances, the greater the atomic

[1] See foot-note, page xxvii of the Introduction.

weight, the less the specific heat; and after examining a number of cases they concluded that it could be stated as a law, that, if the specific heat of a simple substance is multiplied by the weight assigned to the atom of that substance, in every case the product is the same number — 6 ±. This may be seen by inspecting the following table: —

Substance.	Specific Heat.	Atomic Weight.	Product.
Lead	0.0314	207	6.5
Mercury	0.0333	200	6.7
Antimony	0.0508	120	6.1
Silver	0.0560	108	6.0
Zinc	0.0955	65.3	6.2
Iron	0.1138	56	6.4
Phosphorus	0.1887	31	5.8

The explanation by Dulong and Petit for these facts was that *every atom has the same capacity for heat that every other atom has.* The atom of lead weighs 207^{mc}, while that of phosphorus weighs only 31^{mc}; and the capacity of each atom for heat being the same, the amount of heat that will raise 31^{mc} [or grams] of phosphorus one degree, will raise 207^{mc} [or grams] of lead one degree; hence we find a less amount of heat required to raise *one* microcrith [or gram] of lead one degree than to raise *one* microcrith [or gram] of phosphorus one degree. And this is what the above table shows, *i.e.*, that 0.0314 caloric will raise a gram of lead one degree, while it requires 0.1887 calorie to raise a gram of phosphorus one degree.

In other words, as each atom of lead weighs more than each atom of phosphorus, there would be, in equal amounts, — say, ten grams, — of the two substances a less number of lead atoms than of phosphorus atoms, and therefore, *as each atom requires the same amount of heat to heat it up to a given point*, a less amount of heat would be required to heat up the ten grams of lead than the ten grams of phosphorus; or, as we say, the specific heat of lead is less than the specific heat of phosphorus. Examine the table and note that, as the specific heats increase, the atomic weights decrease, and the product of the two is in every case not far from six. In fact, as neither the specific heats nor the atomic weights have been determined with perfect accuracy, it may be supposed that when these determinations shall have been made correctly all the products will be the same.

The discovery of this relation has been of value in this way. Suppose you have found the combining number for copper to be 31.8, and you wish to know whether to take this number 31.8, or 2 times $31.8 = 63.6$, or 3 times $31.8 = 95.4$, or some other multiple of the combining number, for the true weight of the atom. Determine the specific heat of copper, and you will get about 0.09.

$$31.8 \text{ times } 0.09 = 2.862$$
$$63.6 \text{ `` } 0.09 = 5.724$$
$$95.4 \text{ `` } 0.09 = 8.586$$

As 5.724 is so much nearer six than either of the other numbers, the true atomic weight of copper [if the law of Dulong and Petit is to be trusted] is 63.6. After this discovery by Dulong and Petit, it was found advisable to halve the values of several atomic weights used up to that time.

Determine the true atomic weight for zinc from its combining number and its specific heat. In this determination use the combining number and the specific heat number that you found yourself.

If no other method is available, even the atomic weight itself may be determined from the specific heat of an elementary substance; for if 6.4, which is a number very near the average of the products of all reliable atomic weights multiplied by the corresponding specific heats, be divided by the known specific heat, the required atomic weight, or a number very near to it, will be obtained

Though a help in determining the correct atomic weights, perfect reliance must not be placed on this discovery of Dulong and Petit, for there are a few substances, *e.g.*, carbon, boron, and silicon, the products of whose atomic weights multiplied by their specific heats [taken at ordinary temperatures] do not equal 6 ±.

For Review. — § 12. State the second great aid that came to chemists in determining which multiple of the combining number to take for the true atomic weight. Who discovered this aid? When? What is heat supposed to be? Distinguish between temperature and heat. What is the unit amount of heat? What is this unit called? What is a calorimeter? Define specific heat. Describe, briefly, a method for finding the specific heat of a metal. When tables are prepared showing the specific heats and the atomic weights of simple substances what is noticeable? How did Dulong and Petit explain this peculiarity? In what way is use made of the relationship between specific heat and atomic weight? Show how you can tell, by means of the specific heat, the best multiple of the combining number for zinc to choose as the true atomic weight for zinc.

§ 13. Isomorphism.

Still another help in determining atomic weights was found in the discovery of a relation between crystalline form and chemical composition. This discovery was made by a German, Mitscherlich, who announced, just about the same time that Dulong and Petit made their famous discovery, that when two different substances have the same crystalline form their **isomorphism**[1] is due to the fact that the molecules of the two substances have the same number of atoms, and that these atoms are joined in the same way. The *nature* of the atoms was not supposed to make any difference whatever, *i.e.*, one substance might have a chlorine atom where another had a bromine atom, but there must be the *same number* of atoms.

Experiment 32.

Isomorphism.

[a] Take a small amount of *chloride* of sodium; also a little *iodide* of sodium. Recrystallize each from a very strong and *very hot* solution, and when crystalliza-

---

[1] Isomorphism means a similarity of form.

tion has taken place, examine [best under a microscope of low power] the crystals deposited. What is the general form of each?

[*b*] Repeat the experiment, using chloride of *sodium* and chloride of *potassium*.

Let us assume that we know chloride of sodium is made of one atom of chlorine and one atom of sodium; also that we know the total molecular weight of the substance to be 58.5^{mc}— 23^{mc} belonging to the sodium and 35.5^{mc} belonging to the chlorine. Let us also assume that we have found the total *molecular* weight of the iodide of sodium to be 149.9^{mc}. As these two substances are isomorphous, they must, according to the law of Mitscherlich, have the same number of atoms, *i.e.*, the iodide must have one atom of iodine and one of sodium. As the total weight of the two atoms is 149.9, and the sodium weighs 23, iodine must have an atomic weight of $149.9 - 23. = 126.9$.

Assuming that we have found the molecular weight of chloride of potassium to be 74.6, find, by the principle of isomorphism, the atomic weight of potassium.

Though this discovery by Mitscherlich has been a valuable aid, later investigations have shown that it cannot be relied upon, for there are some substances, *e.g.*, sodium sulphate and barium manganate, which crystallize in the same form, but do not have a similar composition.

For Review. — § 13. What was the third great aid that came to chemists in determining which multiple of the combining number for an element should be taken as its true atomic weight? Who discovered this aid? What is isomorphism? Give an illustration. Why can not this method be relied upon?

---

§ 14. Periodic Law.

The last discovery of importance which helps us in determining atomic weights is the law of periodicity. In 1864, Newlands, an English chemist, made an arrangement of the elements according to their atomic weights. He called attention to the fact that, when thus arranged in form of a table, the elements fall into natural groups, each group distinguished by its members having similar properties. Not much serious attention was paid to this table of Newlands. He was even asked, jokingly, if he would not next prepare a table of elements arranged according to the first letters of their names, and see if he could not get similar groups. But, nevertheless, this classification by Newlands contained the germ of an important discovery.

In 1869,[1] Lothar Meyer, a German chemist, published a classification of the elements far more extended and better arranged than Newlands'. He found that if the elements are written in lines from left to right according to their increasing atomic weights, that [excluding hydrogen, the lowest in weight, and beginning with lithium], when the eighth, sodium, is reached, it much resembles lithium, the first; and the ninth resembles the second; and, again, the fifteenth, potassium, resembles lithium, the first; and the sixteenth resembles

---

[1] About the same time Mendeléeff, a Russian chemist, called attention to the fact that he also had come to the conclusion that there was a great law underlying these same facts.

the ninth. Meyer arranged a table showing clearly this *periodic recurrence* of similar properties. There were a number of vacant places in the table, but it was suggested that these might be filled by elements not discovered.

A study of this table shows that elements whose properties are similar may be grouped in natural families, among the members of which there is a regular increase in the atomic weights, and a *corresponding* progressive change in both physical and chemical properties.

Since 1870 much attention has been given to the development of this table, and, in general, all observations have gone to prove that *the properties of any element are periodic functions of its atomic weight;* or, in other words, the properties of elements vary as their atomic weights change. This is called **The Periodic Law,** with which the name of Mendeléeff is so often associated.

To understand the grounds on which the law is based one must have an intimate knowledge of both physical and chemical properties, as well as of the atomic weights, of all the elements — a knowledge that can scarcely be obtained in a year [or perhaps years] of chemical study.[1] Hence we shall make no attempt at doing any experiments to illustrate the periodic law.

The use made of the periodic law in determining atomic weights is as follows: If the properties of an elementary substance are known, the substance can be

---

[1] The author advises a student who can spend a second year on chemistry to devote his second year to Descriptive Chemistry, *i.e.*, largely to a study of properties of substances, rather than to Qualitative Analysis.

fitted into the periodic table among elements of similar properties, and, from its position, its probable atomic weight can be inferred.

It is of interest to note how one of the gaps in the periodic table has been filled. Mendeléeff himself predicted that there was an element [missing] whose atomic weight was 72. From the properties of its neighbors in the table he ventured to predict the properties of this missing element. In 1886 Clemens Winkler discovered a new element whose atomic weight has been determined as 72.3. Let us look at the properties as *predicted* by Mendeléeff and those *found* by Winkler.

PREDICTED.	FOUND.
Atomic weight, 72.	Atomic weight, 72.3.
Specific gravity, 5.5.	Specific gravity, 5.49.
Will form an oxide when heated in the air.	Forms an oxide when heated in the air.
Oxide will have two atoms of oxygen.	Two atoms of oxygen in the oxide.
Easily obtained from its ore by reduction with carbon or sodium.	Easily obtained from its ore by reduction with carbon or hydrogen.
A metal.	A metal.
Dirty grey.	Grey-white.
Will melt with difficulty.	Melts at 900° C.
Will form a chloride with four atoms of chlorine.	Forms a chloride which has four atoms of chlorine.
Chloride will boil near 100°, probably lower.	Chloride boils at 86°.
Will form a sulphide.	Forms a sulphide.
Sulphide will not be soluble in water, but probably will dissolve in sulphide of ammonium.	Sulphide is moderately soluble in water, more readily in sulphide of ammonium.
Scarcely acted on by acids.	Not acted on by acids.

There are at present seventy-two elements recognized. All the aids we have for the determination of their atomic weights have been brought to bear, and many skilful workers have devoted years of time and thought, and are still devoting time and thought, to the accurate determination of the relative weights of the atoms. Still the work is by no means satisfactorily completed. Below is given a list[1] of the seventy-two elements, and the weights now assigned to them. The figures are not given [though in many cases determined] beyond the first place of decimals.

Aluminum,	27.1	Germanium,	72.3	Phosphorus,	31.
Antimony,	120.	Glucinum,	9.1	Platinum,	195.
Arsenic,	75.	Gold,	197.3	Potassium,	39.1
Barium,	137.4	Hydrogen,	1.	Praseodimium,	144.5
Bismuth,	208.	Indium,	113.7	Rhodium,	103.
Boron,	11.	Iodine,	126.9	Rhubidium,	85.4
Bromine,	79.9	Iridium,	193.	Ruthenium,	101.6
Cadmium,	112.2	Iron,	56.	Samarium,	150.?
Caesium,	132.9	Lanthanum,	138.2	Scandium,	44.
Calcium,	40.	Lead,	206.9	Selenium,	79.
Carbon,	12.	Lithium,	7.	Silicon,	28.4
Cerium,	140.2	Magnesium,	24.4	Silver,	107.9
Chlorine,	35.5	Manganese,	55.1	Sodium,	23.
Chromium,	52.1	Mercury,	200.	Strontium,	87.6
Cobalt,	59.	Molybdenum,	96.	Sulphur,	32.1
Columbium,	94.	Neodymium,	141.	Tantalum,	182.5
Copper,	63.6	Nickel,	58.6	Tellurium,	125.
Erbium,	166.?	Nitrogen,	14.	Terbium,	160.?
Fluorine,	19.	Osmium,	190.8	Thallium,	204.2
Gadolinium,	156.1	Oxygen,	16.	Thorium,	233.1
Gallium,	70.	Palladium,	106.6	Thulium,	171.?

[1] Taken from a recent revision by Dr. Richards of Harvard University.

Tin,	119.	Uranium,	240.	Yttrium,	89.?
Titanium,	48.1	Vanadium,	51.3	Zinc,	65.3
Tungsten,	184.	Ytterbium,	173.	Zirconium,	90.6

For Review. — 14. What is the fourth and last great aid that has come to help in the determination of atomic weights? State briefly the history of the discovery of this aid. What is meant by the periodic law? How has this law been used with success?

How many elements are now recognized? Fix in mind the atomic weights now assigned to those atoms in which you feel the most interest.

LANGUAGE OF CHEMISTRY.

The language of chemistry is largely symbolical. It is a kind of shorthand, combinations of letters and figures being used to represent the names of substances, and signs to express processes.

The first letter [or the first and some other prominent letter] of the Latin name of an element is used as the symbol for that element. Thus, H represents an atom of hydrogen; O, an atom of oxygen; S, an atom of sulphur; C, an atom of carbon; Ca, an atom of calcium; Cl, an atom of chlorine. Of course if C has been taken to represent the atom of carbon it cannot also stand for the atom of calcium, hence Ca, the first letter and another prominent letter, are taken for the symbol. In the same way, to represent chlorine, Cl is used.

In most cases the English and the Latin names begin with the same letters. The following are the exceptions:—

ENGLISH.	LATIN.	SYMBOL.
Antimony	Stibium	Sb
Gold	Aurum	Au
Iron	Ferrum	Fe
Lead	Plumbum	Pb
Mercury	Hydrargyrum	Hg
Potassium	Kalium	K
Silver	Argentum	Ag
Sodium	Natrium	Na
Tin	Stannum	Sn
Tungsten	Wolframium	W

LANGUAGE OF CHEMISTRY.

The following is a complete list of the symbols of the seventy-two elements:

Name.	Symbol.	Name.	Symbol.	Name.	Symbol.
Aluminum	Al	Hydrogen	H	Ruthenium.	Ru
Antimony	Sb	Indium	In	Samarium	Sm
Arsenic	As	Iodine	I	Scandium	Sc
Barium	Ba	Iridium	Ir	Selenium	Se
Bismuth	Bi	Iron	Fe	Silicon	Si
Boron	B	Lanthanum	La	Silver	Ag
Bromine	Br	Lead	Pb	Sodium	Na
Cadmium	Cd	Lithium	Li	Strontium	Sr
Caesium	Cs	Magnesium	Mg	Sulphur	S
Calcium	Ca	Manganese	Mn	Tantalum	Ta
Carbon	C	Mercury	Hg	Tellurium	Te
Cerium	Ce	Molybdenum	Mo	Terbium	Tb
Chlorine	Cl	Neodymium	Nd	Thallium	Tl
Chromium	Cr	Nickel	Ni	Thorium	Th
Cobalt	Co	Nitrogen	N	Thulium	Tu
Columbium	Cb	Osmium	Os	Tin	Sn
Copper	Cu	Oxygen	O	Titanium	Ti
Erbium	Er	Palladium	Pd	Tungsten	W
Fluorine	F	Phosphorus	P	Uranium	U
Gadolinium	Gd	Platinum	Pt	Vanadium	V
Gallium	Ga	Potassium	K	Ytterbium	Yb
Germanium	Ge	Praseodymium,	Pr	Yttrium	Yt
Glucinum	Gl	Rhodium	Rh	Zinc	Zn
Gold	Au	Rhubidium	Rb	Zirconium	Zr

The atomic weight of an element should always be associated with its symbol, thus, H should represent to your mind 1^{mc} of hydrogen; C, 12^{mc} of carbon; Fe, 56^{mc} of iron; Zn, 65.3^{mc} of zinc, and so on.

To express compounds there are used symbols called formulae, made by writing together the symbols of the atoms that are in the molecule of the compound, thus,

HCl is the formula for a molecule of hydrochloric acid; NaOH represents a molecule of hydroxide of sodium. When there are several atoms of the same kind in the molecule subnumerals are used, thus, H_2O stands for a molecule of water; HNO_3, for one of nitric acid; H_2, for a *molecule* of hydrogen gas; O_2, for a molecule of oxygen gas; $CaCO_3$, for one of carbonate of calcium; NH_3, for one of ammonia; and H_2SO_4, for one of sulphuric acid.

If the symbol for every atom has the correct atomic weight associated with it, there is no difficulty in telling the molecular weight of a substance if its formula is known, for the molecular weight must be the sum of the weights of all the atoms that go to make up the molecule. What, then, is the molecular weight of hydrochloric acid? Of carbonate of calcium? Of sulphuric acid?

The formula of a molecule always tells three things: first, what the substance is; second, of what atoms it is composed; third, what the molecular weight is.

To express two or more molecules, or atoms, coefficients are used. Thus, 2 H_2O represents two molecules of water; 5 H_2SO_4 stands for five molecules of sulphuric acid; 10 K_2SO_4, for ten molecules of sulphate of potassium; 2 H, for two *atoms* of hydrogen; 2 H_2, for two *molecules* of hydrogen; 7 NaCl, for seven molecules of common salt; 3 O, for three atoms of oxygen; 3 O_2, for three molecules of oxygen; and 3 O_3, for three molecules of ozone—a substance we have not studied.

Note. Be sure that you see the distinction between atomic and molecular expressions. State which of the following are atomic and which molecular: H, H_2, 7 H, 2 H, 5 Cl_2, HCl, H_2O, H_2O_2, 3 O.

In chemistry the action of one substance on another is represented by the sign $+$, and equations are used to express chemical changes, the factors of the change being put before the sign $=$, and the products after. Thus, $ZnO + H_2SO_4 = ZnSO_4 + H_2O$, represents the change that takes place when the oxide of zinc acts on sulphuric acid. When water is used simply to produce solution, the symbol Aq, standing for the Latin word aqua, is used. Thus the reaction of ZnO and H_2SO_4, as we conducted it, would be represented as follows: $ZnO + [H_2SO_4 + Aq] = [ZnSO_4 + H_2O + Aq]$. The brackets are used to indicate the solutions. As the Aq seems to play no part chemically in the changes, this symbol is frequently omitted altogether. The neutralization of hydrochloric acid with hydroxide of potassium is represented thus: $HCl + KOH = KCl + H_2O$; and the neutralization of sulphuric acid with hydroxide of sodium thus: $H_2SO_4 + 2\ NaOH = Na_2SO_4 + 2\ H_2O$.

Write in equation form the changes that were brought about:

[a] When an atom of zinc acted on a molecule of sulphuric acid in water solution;

[b] When oxygen and hydrogen gases were produced from water by the electric current. It will not do to write this $H_2O = H_2 + O$, because *molecules* of oxygen gas came bubbling up. The equation should be written, $2\ H_2O = 2\ H_2 + O_2$;

[c] When oxygen was made by heating the chlorate

of potassium. Here assume, as in the last case, that the gas came off in molecules. The formula for a molecule of chlorate of potassium is $KClO_3$, and that for the chloride which results is KCl;

[d] When oxygen gas acted on phosphorus. The formula for a molecule of phosphorus is P_4, and that for the resulting white oxide is P_2O_5;

[e] When an atom of carbon burned in oxygen, producing the dioxide of carbon, CO_2;

[f] When an atom of sulphur burned in oxygen, producing the dioxide of sulphur, SO_2;

[g] When a molecule of water was added to a molecule of dioxide of sulphur, and there resulted a molecule of sulphurous acid, H_2SO_3;

[h] When O_2 and SO_2 were passed over hot platinum sponge, and SO_3—the second oxide of sulphur—resulted;

[i] When sulphuric acid—H_2SO_4—was produced by the addition of water to the second oxide of sulphur;

[j] When two molecules of HCl, in water solution, reacted with one molecule of carbonate of calcium, $CaCO_3$;

[k] When five volumes of H_2 acted with two volumes of NO, and there resulted two volumes of ammonia—NH_3—and two volumes of steam.

Note that these chemical equations are nothing more than a short way of expressing chemical facts. They resemble algebraic equations, inasmuch as there are the same atoms on one side of the sign of equality that there are on the other side; and, consequently, the sum of the atomic weights on one side equals the sum of those on the other. But it is not true that any transposition that can be made in algebra can be made in a chemical equation. Chemical equations are far less flexible than algebraic ones, the chemical being limited to the expression of observed facts. This is an important point, and should not be forgotten.

A *chemical equation* always expresses two of the great laws of chemistry, and, when gaseous substances are involved, it expresses a third. It always expresses the law of conservation of mass, as is shown by the use of the equality sign. It also always expresses the law of definite proportions by weight, because the symbols used stand for definite weights of matter. And, when any of the substances are gases, the simple ratio between their coefficients expresses the law of definite proportions by volume.

Note, that, in addition to the three facts [see page 220] that the *formula for a molecule* always expresses, there is a fourth shown when the formula is that of a gaseous substance. The simple molecular formulae of gases all represent the same volume. O_2, H_2, Cl_2, N_2, H_2O [steam], NH_3, NO, NO_2, CO_2, SO_2, CO, and HCl all represent the same volume; for, according to the hypothesis of Avogadro that in equal volumes of all gases there are the same number of molecules, it follows that every molecule occupies just as much space as every other molecule. Hence the formula for the molecule of any gaseous substance represents a definite volume of that substance.

STOICHIOMETRY.

If we know the equation that expresses any chemical change and know the weight in grams of a single substance that enters into the change we can calculate the weight of every other substance involved. This power to calculate unknown weights of substances is one of great value to the chemist. If, for instance, he is called upon to produce 100 grams of silver chloride, it is not necessary for him, as it was in the old days, to guess at the amounts of nitrate of silver and of common salt to use, with a good chance of making just a little less than the ten grams or a considerable amount more, nor need he go "by rule of thumb." By the aid of a chemical equation and his knowledge of the atomic weights he can calculate the exact amounts of the factors required to yield the required weight of product. Let us make the calculation. First write the equation expressing the change. It is this: $AgNO_3 + NaCl = AgCl + NaNO_3$. Now write *above* the formula for each molecule its molecular weight, *i.e.*, the sum of the weights of all the atoms in the molecule. As the weights, in grams [or in pounds, or in terms of any other unit], bear to each other the same ratio as do the molecular weights, proportions can be made enabling us to find the unknown amounts, *e.g.*, the molecular weight of the AgCl is to the molecular weight of the $AgNO_3$, as is the gram weight of AgCl [known to be 10] to the gram weight of $AgNO_3$, or $143.5 : 170 :: 10 : x$. $x = 11.85 -$. Therefore 11.85^g of nitrate of silver will be required to produce 10ᵍ of chloride of silver. Now calculate how much NaCl is needed. Also calculate the weight of the nitrate of sodium that will also result from the change. See if the sum of the weights of the products equals the sum of the weights of the factors. This finding of unknown amounts is called **stoichiometry.**

The chief rule of stoichiometry is as follows: When the gram weight of one substance which is involved in a chemical change is known, in order to find the gram weight of any other substance involved, make a proportion thus: *the molecular weight of the known is to the molecular weight of the unknown as the gram weight of the known is to x.* x will be the required gram weight.

If it is desired to find the *volume* of a gas that will be produced in a given case, first find the *weight* of gas produced: then calculate [making use of the known density of the given gas] the volume this weight would occupy at N. T. P.: and, finally, calculate [using the laws of Boyle and of Dalton] the volume which this volume at N. T. P. will occupy under the given conditions.

MANIPULATIONS.

The following directions may be of service in performing the mechanical operations that so often occur in laboratory practice:

To Mark Glass.

This can best be done with a pencil that has recently been invented for the purpose. Each student should keep one of these pencils constantly at hand in his desk. By its frequent use, particularly in marking the point at which tubes are to be cut, and in labelling vessels that have to be set away at the end of laboratory periods, much time and annoyance will be saved. These pencils may be obtained from the publishers of this book — Ginn and Company.

To Cut Glass.

Small and medium tubes, and rods, are best cut as follows: With a sharp triangular file make a short scratch at the point where the tube is to be cut. Take the tube in your hands with the two thumbs on the tube *under* the scratch, and the fingers of each hand spread out on the tube to the right and the left of the scratch. The nails of the two thumbs should be close together and touching the glass exactly under the scratch. By means of the two little fingers bring a gradually increasing downward pressure to bear, letting the thumb nails act as a fulcrum.

Larger tubes, *e.g.*, combustion tubes one or more cm in diameter, may be cut in a similar manner by substituting the edge of a triangular file, resting on the desk, for the fulcrum, and pressing downward with the palms of the two hands. If, however, the tube to be cut is very short, proceed thus: Make the scratch with the file, then either touch this scratch suddenly with a hot iron, *e.g.*, the end of an old file heated red-hot in the Bunsen flame; or drop upon it a bead of red-hot molten glass; or, best of all, apply to it a small gas-flame, which may be made by disconnecting the Bunsen burner from its tube and inserting in the end of the tube a tip of *hard* glass drawn out like the wash-bottle tip. The best flame is a vigorous, clear-cut, smokeless one, about 1^{cm} long, produced by a small opening and a full pressure of gas. When such a flame is applied to the scratch the tube usually snaps square off. If it does not snap within half a minute, the change of temperature has not been sudden enough. Cool the tube, and apply the flame again. If a crack starts but does not run entirely around, or does not run straight, turn off the gas till the flame is only about 2^{mm} long, and apply this flame to the glass just beyond the end of the crack. The crack will advance to the flame. Move the flame ahead, and in this way lead the crack around wherever you wish.

To cut off the necks of bottles, beakers, and test-tubes, make a scratch with a file, and start a crack as directed above. Lead this crack, best with the 2^{mm} flame, around the neck of the vessel. In this way chipped or cracked beakers and test-tubes may often be made into serviceable articles, by cutting off their tops and fire-polishing the sharp edges.

To Fire-polish the Edges of Glassware.

All sharp glass edges, *e.g.*, ends of tubes and rods, freshly cut necks of bottles, broken edges of beakers and test-tubes, should be melted in the Bunsen flame till they are smooth and rounded. The flame of the Bunsen burner is usually sufficient, but the work can be done quicker in the flame of the blast-lamp. Great care must be taken, particularly with the latter flame, not to crack the glass by too sudden application of the heat. Begin with a small flame, which must be increased slowly. When heating glass there is less danger of its snapping if it is immersed first in the less vigorous, luminous, sooty flame, and heated as hot as possible there before it is put in the Bunsen flame. In the case of thick glass the article should not at first be held continuously, even in the luminous flame, but should be immersed for a moment, then withdrawn for a moment, again immersed, and so on, till it is thoroughly heated. Thick glass is more apt than thin to crack by a sudden application of heat, because the side where the heat is applied expands before the heat has passed to the other side and expanded that,— hence a rupture.

To Bend Glass Tubes and Rods.

This must be done in a wide flame, and not in the common Bunsen burner flame, and, except in the case of very short tubes, not in the blast-lamp flame. A common bat-wing gas jet [used for illuminating purposes] is good, but leaves the tube sooty. Better than this is the broad tip furnished for this purpose with some Bunsen burners.

Hold the tube, at the point where it is to be bent, in the hottest part of the broad flame. In the case of an ordinary tube, a length of at least 6^{cm} should be immersed in the flame. Turn the tube slowly, so that all sides shall be equally heated. When the tube has softened, and you feel it to be flexible, *remove it from the flame*, and quickly, but deliberately, bend it to the exact angle wanted. If the right angle is not obtained at the first attempt, it is hardly worth while to heat again and try to bend it into shape. Better take a new piece of glass and start again.

To Draw Out Glass Tubes.

Begin as directed for bending tubes, but heat more. When the glass has softened so much that it begins to sag, remove it from the flame and draw it to the shape wished. Wash-bottle tips, and the like, may be made either with the bat-wing flame or in the Bunsen burner flame. Large or very hard tubes best be drawn with the aid of the blast-lamp. In using the blast-lamp for this purpose care should be taken to apply the heat gradually, as directed under fire-polishing. When it is desired to draw out a long length of capillary tube, the glass should be made very soft, removed from the flame, drawn to the right diameter, *allowed to cool a few seconds in order to take a "set,"* and then rapidly drawn to obtain length.

To Make a Matrass.

A matrass is a hard glass vessel with a long slim neck and a bulb-like body. For a large matrass a Kjeldahl flask serves excellently. Small matrasses, often called "bulb-tubes," are frequently used when a substance has to be heated in a flame so hot that it would soften an ordinary test-tube.

To make a small matrass: Take a piece of hard glass tube of 5–10^{mm} bore. Hold one end in the blast-lamp flame till the glass is soft. With the forceps pinch the softened walls together, and quickly draw off, and reject, about 1^{cm} of the end of the tube. Immerse the closed end of the tube in the blast-lamp flame; hold it there till a lump of softened glass has collected; then put the open end in your mouth and blow gently till the lump of glass forms a bulb.

To Render Corks Air-tight.[1]

Melt some solid paraffine — a candle will do — in a dipper. Roll the cork a little under the foot, and then soak it for a few minutes in the melted paraffine. Never attempt to tighten leaky joints by applying paraffine, sealing-wax, etc. This application is not worth the trouble, seldom stops the leak, and looks shiftless, as apparatus properly constructed never needs any such doctoring.

---

[1] Good corks need no treatment, but in a gross of corks it is seldom that there are not a number which are not air-tight.

To Render Joints Air-tight.

Before putting stoppers in the necks of bottles, rubber bands on jars, rubber hose on tubes, etc., the surfaces of contact should be greased, if an air-tight joint is wanted. Vaseline makes an excellent grease.

To Cut Rubber Neatly and Quickly.

Use a sharp knife, and *keep the cut wet*. A small oil-stone should be kept at hand in the laboratory for sharpening knives.

To Pass a Glass Tube Through a Hole in a Rubber Stopper.

Apply vaseline or glycerine to the glass. In this way a tube that seems too big for the hole may be slipped through easily, particularly if its end has been fire-polished.

To Bore a Round Hole in Glass.

It is often desirable to make a hole in the side of a bottle, test-tube or beaker, without cracking the surrounding glass. Take a triangular file [an old one will do], break off its tip so that a jagged point is made. *Keep this point constantly wet with vaseline* or some other greasy substance, and with a circular grinding motion of the arm and hand, bore into the glass. As the file point gets blunted break off farther down. Do not apply much pressure in the case of beakers and thin glass, but considerable in the case of stout bottles.

When once the file has gone through, take a rat-tail file and, at first *cautiously*, file the hole to the desired size. Do not hold the glass in such a position that the hand would be cut in case there should be a sudden collapse of the article on which you are operating.

To Prevent Mixing Glass Stoppers.

Never take the stopper from a laboratory bottle and lay it on the desk. Always take it out by means of the middle finger and the forefinger with the palm of the hand held upward. This way enables you to hold the stopper [inverted] between the fingers while the hand is free to lift the bottle and pour its contents. Stoppers left on the desk get contaminated and mixed.

To Hold Hot Beakers, Test-Tubes, etc.

A beaker of boiling water may be lifted from the fire if the fingers grasp it by the rim only. A piece of paper, folded into a band 15–20^{cm} long and about 1^{cm} wide, passed around the neck of the test-tube so that the fingers may grasp the two ends of the paper, forms as convenient a holder for a test-tube in which you are boiling a liquid as any of the fancy ones you can buy.

To Use the Pneumatic Trough.

Note. A small iron sink can well be converted into a pneumatic trough in the manner described in Appendix P. Every pneumatic trough should have a "bridge." The bridge recommended for the sink [Appendix P] will do for any trough.

For most purposes the trough should be filled with cold water. For collecting gases that are soluble in cold water, *e.g.*, laughing gas, hot water serves well. In those cases in which it is desired to collect gases free from water vapor, mercury is generally the best liquid to use in the trough. In fact mercury is an excellent substance to use at almost all times, and it is unfortunate that its cost prevents its more general use. As mercury amalgamates with zinc an iron [or porcelain] trough is to be preferred to one of zinc.

When a gas is to be collected by means of the pneumatic trough, the liquid in the trough should reach well above the bridge; the end of the delivery tube should be directly under the hole in the bridge; and the receiving vessel, inverted and filled with the liquid of the trough, should stand on the bridge directly over the hole.

To Use Filter Papers.

The filter paper should be cut in circular form. For most purposes the paper should be folded together so as to halve its surface; then it should again be folded together, thus quartering it; finally, it should be opened in such a manner that three of the quarters are together and form one half of the sides of a cone, while the fourth quarter forms the other half. In fitting the filter to the funnel, the tip of the cone should be inserted in the point of the funnel, and the paper slightly wet by means of the wash-bottle [unless water would harm the substance to be filtered]. In general, the filter paper should be of such diameter that it does not reach, when in place, quite as high as the edge of the funnel.

For substances that filter slowly a plaited filter is useful. To make a plaited filter: fold to quarters, as in the case of the common filter; then open to a half filter; the half will show a fold dividing it in halves; divide each of these halves in halves by a *similar* fold. Again open out to a half filter. Be sure that the three creases which divide the half filter into quarters are all alike, *i.e.*, all have their ridges on one side and hollows on the opposite. Next make a series of creases down the middle of each slice, taking care to have these latter creases point exactly opposite from the first. In making the creases be careful and do not tear the tip of the cone. Now open the paper completely, and insert it in the funnel in such a way that nowhere shall there be two thicknesses together. The plaits of this funnel furnish a large amount of surface for filtration.

To Dry Bottles, Flasks, etc.

Fit a long, hard glass tube to the rubber tube from the bellows. Hold the glass tube in the flame of a Bunsen burner and force a stream of hot air into the article that is to be dried. Direct the stream of hot air against any drops of moisture that you may see.

To Remove Stoppers that have Stuck.

First gently tap the stopper against some rigid body — as a brick wall or iron beam. If this treatment fails to start the stopper, immerse the bottle, top down, in a vessel of water. Allow the stopper to remain immersed for several hours. If this fails to start the stopper, wipe the bottle dry and apply to its neck the heat from the small gas flame mentioned under "To Cut Glass," page 227. Turn the neck as you apply the flame.

To Pour Gases.

With a little care gases may be poured in a manner similar to that in which you pour liquids. Of course a gas lighter than air must be poured *up* into an inverted vessel. In pouring a gas from one tt to another it is well to hold the fingers and hand around the mouths of the two tt's in a way to protect the stream of gas from any current of air that might blow it to one side.

To Use a Bunsen Burner.

For almost all work the Bunsen burner should be used with its air-vent open. When this vent is open the air enters and mingles with the gas. Then when the gas issues from the top of the burner it burns rapidly, with a very hot and almost colorless flame.

If for any reason you are not using the full supply of gas, the supply of air should be reduced to a corresponding amount by partly closing the vent.

Sometimes a sudden gust of wind will drive the flame down the tube of the Bunsen burner and ignite the gas as it issues below. In this case the burner is said to "snap," or "strike back." A "snapped" burner can usually be detected by the peculiar color of the flame it shoots up; also by a peculiar odor which fills the atmosphere all around the burner. The danger from a "snapped" burner is that the base gets intensely hot and is apt to burn the fingers. A "snapped" burner should have its supply of gas wholly turned off and be re-lighted.

To Use the Bunsen Blast-Lamp.

Allow the gas to circulate in the outer chamber, and force a current of air in through the small tube in the center.

APPENDICES.

APPENDICES.

APPENDIX A.

Apparatus for the Electrolytic Decomposition of Water.

Prepare a vessel for holding water as follows: Take any common bottle of about one quart capacity. At a point about one-third down from the neck to the bottom cut[1] off the top part parallel to the bottom, so that the top part, inverted and with a cork in its neck, will form a stout, shallow dish. If the neck is a long one, cut it off so that not more than one inch is left. Fit a cork to the neck. Pass two platinum wires, each about 15^{cm} long, between the cork and the glass, well up into the body of the vessel. About one inch of wire should project out from the neck. The wires should be as far apart from one another as possible, and the parts [about 10^{cm} long] that protrude into the vessel should be twisted into spirals. At their lower ends the platinum wires should be connected, by being tightly twisted, or, better, soldered, with copper wires leading to two or more Bunsen cells or other source of electricity. Support the vessel, with the neck down, on a ring of the ring stand. Light a candle, and, holding the candle

---

[1] For cutting thick glass, see **Manipulations**.

inverted, let the candle grease drop down and fill the neck of the bottle above the cork while you hold the wires apart and keep their spirals well above the grease. The grease makes the joints water-tight.

For the Bunsen cell have ready a two-quart glass battery jar, a porous cup as tall as the battery jar, a battery zinc to go around the porous cup, a battery carbon to go in the porous cup, two [one foot long] pieces of copper wire, and two binding screws to connect wires with the zinc and with the carbon. Fill the jar nearly half full of water. Add about one-tenth as much strong sulphuric acid as you have added water. Add the acid slowly, with constant stirring with a glass rod. **Caution:** *Sulphuric acid is very corrosive. Do not get any on skin or clothes.* Put the zinc in the jar of acidified water, and let the acid work for about a minute. Invert the zinc, and let the acid again act, now on the upper part, for about a minute. Then remove the zinc, and, at the sink, with a rag and a little mercury, rub mercury into the zinc, both inside and outside the cylinder, till the surface is bright and well amalgamated. Amalgamation prevents the zinc from being unduly eaten away by the acid. Put the zinc back, right side up, in the acid. Put the porous cup within the zinc. If too much acid has been put in the outer glass vessel, now remove some. Fill the porous cup about half with nitric acid, and the other half with sulphuric acid, taking care that the level of the acid in the porous cup is the same as that of the liquid in the glass vessel without the porous cup. Put the carbon in the porous cup. By means of the binding screws and a short piece of copper wire, connect the carbon of the cell with one of the platinum wires, and connect the zinc with the carbon of another cell. Be sure that all points of contact between the wires and binding screws have been freshly scraped clean and bright.

Note. Two cells are necessary. More than two will cause the decomposition of the water to be more rapid, and, consequently, more satisfactory.

Connect the zinc of the second [or last] cell with the other platinum wire of the decomposition apparatus.

APPENDIX B.

Hydrogen Explosions.

One of the most frequent of accidents in a laboratory for elementary chemistry is the explosion of a mixture of hydrogen and air. When generating hydrogen which is to be lighted or near which any kind of fire is to come, always test its explosive qualities, *i.e.*, test to see whether air is still mixed with the hydrogen.

The test is best made by catching a small tt full of the gas, and [having removed the tt, while still inverted, quickly, several feet from the supply of hydrogen] touching a match to the open inverted mouth of the tube. A sharp ringing report shows danger, *i.e.*, that there is an explosive mixture of air and hydrogen present. Continue testing till, when the flame is applied, there is only a gentle pop, and the hydrogen burns quietly with a faint flame up into the tt.

A tt of about 1^{cm} bore, and not more than 6 or 8^{cm} long, is the best for this test. Such a tube may be made from ordinary large soft glass tube or by cutting a common small tt in halves. Hydrogen explosions are dangerous from the flying glass that is usually sent in all directions. This method of testing is called "by the explosion tube."

APPENDIX C.

Test Papers.

Note. The only test papers that are needed in elementary chemistry are: one to indicate an acid solution, and one to indicate an alkaline solution. In preparing such test papers use is made of the changes of color produced by acids and alkalies on certain vegetable substances. Litmus in acid solution is red, and in alkaline, blue.

Take 1 part [*e.g.*, 10^g] of the litmus of trade. Mix it in a porcelain evaporating dish with 6 parts [*e.g.*, 60^g] of water. Warm gently, and stir for about ten minutes. Filter. Take half the filtrate, and, with a glass rod, stir in a drop or so of very dilute sulphuric acid [*e.g.*, 1^{cc} of acid to 50^{cc} of water]. Stir in acid till the solution is *just* turned red. Immerse strips of filter paper in the red solution, keep the solution warm for a little while, remove the strips and hang them on a line [in a room free from ammonia] to dry. They should be of a good pink color.

Take the remainder of the red solution and add some of the blue [previously saved apart], till the color is again turned blue. Immerse strips of paper in this liquid and hang them to dry in a room free from acid fumes. To detect an acid use the blue paper; to detect an alkali use the red. Never put a test paper in any solution, but always take out a drop on a glass rod, and touch this drop to the test paper held between the fingers. Should the paper be put in the solution, the coloring matter would contaminate the solution itself. It is never safe to lay a test paper on a desk, for the desk itself may contaminate the test paper.

For detecting alkaline substances papers treated with turmeric are in some respects preferable to those treated with red litmus. To prepare turmeric papers, take 2 parts of water and 4 of *alcohol* to 1 of turmeric. Warm *gently*, and stir for about ten minutes. Best warm the alcohol solution over hot water, not over a flame, for fear of fire. Filter. Immerse strips of filter paper in the liquid. Keep the papers in the solution [which should be kept warm] for a few minutes, and hang on a line to dry as in the preparation of litmus paper.

APPENDIX D.

Suction Pumps.

Every laboratory should possess some form of suction pump, both for general use and, in particular, for rapid filtrations. The Bunsen filter pump, though expensive [$7–$10], is excellent where there is no water pressure available. This pump is made on the principle of the Sprengel mercury pump. Where water under pressure can be had any one of the satisfactory cheaper pumps, as the Richards, or the Fischer [cost $1–$3], may be used.

For rapid filtrations use a *stout* glass bottle as a receiver for the filtrate. This bottle should be fitted with a two-hole *rubber* stopper. Through one hole of the stopper pass the stem of the funnel; through the second hole pass a piece of glass tubing, which can be connected by a stout rubber hose with the filter pump. Make [from a round piece of platinum foil about one inch in diameter] a little platinum cone to fit the glass funnel. Use this cone to protect the tip of the filter paper when the suction pump is used. In filtering, allow the filter pump to exhaust the air from the flask. The pressure of the atmosphere then presses the liquid rapidly down into the flask.

APPENDIX E.

Catch-Bottles.

It frequently becomes necessary to "wash," or purify, a gas. For this purpose the gas is made to pass through water, sulphuric acid, or some other liquid, contained in a "catch-bottle." As it is usually necessary to dry the gas after it has been washed, it is well to have on hand two catch-bottles connected and arranged so that at a moment's notice the wash-liquid may be put in one and sulphuric acid in the other.

Take two 2-oz. saltmouth bottles. Fit each with a good two-hole cork, or, better, with a two-hole rubber stopper. Pass a glass tube through one hole of the stopper of the first bottle down to the very bottom. This tube should be bent at a right angle just above the cork, and to it should be attached the tube that is delivering the gas which is to be purified. Through the second hole of this first bottle pass another glass tube, beginning just at the lower surface of the stopper and extending over to the second bottle and passing through one hole of the stopper of the latter and down to the very bottom. The two bottles need not stand more than an inch or two from each other. Through the second hole in the stopper to the second bottle pass another short piece of glass tube [bent at a right angle] just through the stopper. This last is the exit tube for the gas after it has been washed and dried.

Put the wash-liquid in the first bottle, and fill the second about half full of sulphuric acid. After the gas has come in through the first tube of the first bottle it should pass down under the liquid and bubble up; then it should pass through the tube to the second bottle, where it should bubble up through the sulphuric acid and pass out through the short exit tube. *Be sure all tubes are arranged properly*, and that all joints are tight before you pass any gas through the bottles.

APPENDIX F.

Generator for Gases.

Take a common narrow-necked bottle of 8 to 10-oz. capacity. Bore[1] one or two holes of 4 or 5^{mm} diameter in the bottom of the bottle. Fit the bottle with a good cork [better, a one-hole *rubber* stopper] and a short glass delivery tube bent at a right angle and ending in a short piece of rubber tube carrying a screw pinch-cock, by which the flow of the gas may be regulated. Put a layer of broken glass about one inch deep in the bottom of the bottle, and set the bottle itself in a large beaker [or, better, in a battery jar, or in a vessel made from a large bottle whose top part has been cut off].

To Generate Non-Combustible Oxide of Carbon. Put small lumps of hard marble in the generator. Insert the stopper, close the pinch-cock, and pour common muriatic acid in the outer vessel till the surface of the liquid approaches within about one inch of the top edge. Open the pinch-cock and the acid will force its way into the bottle and act on the marble. When the pinch-cock is closed the accumulating gas forces the acid down and out, and the action ceases. The broken glass serves for draining the marble, and causes the action to cease more promptly than it would were the marble placed on the bottom. After the gas has been generated it should be washed and dried by being passed through two catch-bottles, the first of which contains water and the second sul-

[1] See Manipulations.

phuric acid. The water serves to stop the passage of any hydrochloric acid that may be driven over, and the sulphuric acid removes the water with which the gas at first is always charged.

To Generate Hydrogen Sulphide. Put sulphide of iron instead of marble in the generator. The sulphide should be broken in small pieces. Sulphuric acid somewhat diluted [five volumes of water to two volumes of acid] may be used with advantage in the outer vessel.

To Generate Hydrogen. Put zinc in the generator, and dilute sulphuric acid [five volumes of water to one of acid] in the outer vessel.

APPENDIX G.

Hood.

There should be in the laboratory a small closet provided with a flue leading up some chimney or into the open air. This closet should have a glass window that can be opened and closed so that articles put "under the hood" can be shut off from the main room. It is well, when possible, to have an artificial draught produced by a steam coil, burning gas jet, or the like; also to have gas, water, and a sink under the hood.

APPENDIX H.

Preparation of Chlorine.

Fill a small flask about one quarter with small *lumps* of black oxide of manganese. Add common muriatic acid till the liquid and the black oxide together fill about one half of the flask. Insert a one-hole cork and delivery tube. Set the flask with its charge in a beaker, or a pot, of water. Heat the water, and chlorine gas will pass off. The delivery tube should be connected with two catch-bottles,[1] the first containing water, the second sulphuric acid. The first catch-bottle serves to catch any hydrochloric acid that is carried over, and the second to dry the chlorine.

APPENDIX I.

Sodium Amalgam.

For the preparation of sodium amalgam on the large scale, have ready a large Hessian crucible capable of holding at least ten times as much amalgam as it is your intention to make. Put the mercury in the crucible, and heat with a Bunsen burner. Take the temperature of the mercury by means of a thermometer. Have ready one tenth, by weight, as much sodium as there is mercury. The sodium should be in a single piece. When the temperature of the mercury has reached 200° remove the burner, drop the sodium in the hot mercury, cover the crucible with an iron plate, or with the bottom of some old iron pan, and *at once step back.* The union takes place with considerable commotion. Before the molten amalgam has time to solidify, pour it out in a thin layer on some smooth surface that will not be harmed by the heat. Just as soon as the amalgam is cool enough to handle, break

[1] See Appendix E.

it in small pieces, and store it in a tightly-stoppered bottle that the air may not give it a troublesome coating of hydroxide.

Amalgam that has been spoiled by standing long in the air should not be thrown away, but should be treated with water that the mercury may be recovered.

APPENDIX J.

Test Solutions.

Caution! *Do not get any of either solution* [1] *or solution* [2] *in the mouth.*

[1] Preparation of an Arsenical Solution for Testing.

Weigh out exactly 0.01^g of the white oxide of arsenic, put this in a 250^{cc} flask, add 250^{cc} of water and about a gram of sodium hydroxide, and shake till solution takes place. Use only a single cc of this solution at a time, as the test is a very delicate one, and the arsenide of hydrogen formed is a most terrible and deadly poison.

In adding the solution through the funnel-tube care should be taken to let the solution trickle down the tube in such a way that no air is carried into the bottle. Why avoid letting air enter?

[2] Preparation of an Antimonial Solution for Testing.

Weigh out 0.03^g of tartar emetic [a compound which contains about 40% of antimony]. Dissolve in 250^{cc} of water and use one cc at a time, as in the case of the arsenical solution.

Caution! *Do not get any of either of these solutions in the mouth.*

APPENDIX K.

Use of the Mouth Blow-Pipe.

Rest the smaller end of the blow-pipe on the edge of the Bunsen burner's top, and blow directly through the flame. The gas will be carried along with the blast, and a slender cone of blue flame will be projected at a right angle to the original direction of the flame. If the blast is not powerful enough to use all the gas, turn off the supply of gas somewhat. This blast flame, though small, will be found to be intensely hot. By tipping the Bunsen burner you can direct the flame in any direction.

In using the blow-pipe, make a reservoir of your mouth, while you breathe entirely through the nose. With a little experience you will be able to blow so steadily that the small flame will show little or no fluctuation for minutes at a time.

When it is desired to oxidize a substance care should be taken that only the outer part of the blast flame touches the substance. The inner part of the flame is made up largely of unconsumed gas which, in its eagerness to get oxygen, acts as a reducer. Hence the inner part of the flame is often used to bring about reductions on a small scale.

APPENDIX L.

Arsenical and Antimonial Papers for Testing.

For the arsenical paper it is best, when possible, to get a piece of wall paper that is known to contain arsenic. If this cannot be found, take a piece of filter paper, and dip it in an arsenical solution made five times as strong as that of Appendix J [1]. Hang on a line to dry. **Caution!** *Do not get poisoned.*

The paper poisoned with antimony may be made by dipping a piece of filter paper in a solution five times as strong as that of Appendix J [2].

APPENDIX M.

To Dry Precipitates.

Have ready a clean porcelain evaporating dish. Set this dish on gauze above the flame of a Bunsen burner that has been turned down to only one fourth of its ordinary size. Carefully remove the precipitate, paper and all, from the funnel, and place it in the evaporating dish. Watch the filter paper, and if it shows the least sign of charring, turn the flame still lower.

If you are in no hurry, a good way is to leave the precipitate in the funnel and set funnel and all into a ring supported just above an iron plate which is heated by the Bunsen burner.

Excellent little ovens are sold for the purpose of drying precipitates and the like, but these are somewhat expensive.

APPENDIX N.

To Nurse a Crystal.

After a first crystallization has taken place, select the best formed crystal, remove it from the "mother liquor" [as the liquid in which the crystals were formed is called], and set it aside on a bit of filter paper. Then dissolve the rest of the crystals, by heat, in the mother liquor, and add a *little* more water. Filter, if dirty. Cool the solution. Immerse the reserved crystal in the solution, and set aside till another deposit forms. Much of this second deposit will take place on the large crystal which is thus made to grow at the expense of the smaller. Again remove the large crystal, and repeat the process. Repeat as many times as desired.

It is well, in the beginning, to save *two* or *three* crystals and nurse them together, as nursing is apt to distort a crystal, particularly when the crystal lies each time on the same face. Hence, it is well, also, to turn the crystal occasionally. If you start with two or three crystals, you improve your chances of getting a single good one in the end. Of course, any crystal that has become badly distorted best be dissolved, that it may furnish material for the growth of others.

APPENDIX O.

Distilled Water.

There are but few experiments in this book that demand the use of distilled water. For those few the water may be purchased from the nearest apothecary shop, or it may be distilled by the student, as follows: Take a glass retort [or a "boiling flask"] and fill it half or two thirds with water from the tap. Any retort that holds 250–500^{cc}, or even the wash-bottle flask fitted with a one-hole cork and delivery tube will do. Support the retort on a stand in such a way that the water may be heated by the Bunsen burner. Heat the water rapidly till it boils. Keep it boiling gently. The steam that passes down the neck of the retort will be condensed and drip down in drops of nearly pure water, while the greater part of the impurities will be left behind in the retort. If a boiling flask is used, it is best to pass the steam on through a piece of glass tube in order that more may be cooled and condensed. Do not boil off more than three quarters of the total amount of the water, as there is danger of impurities coming over in the last quarter. It is also well to reject the *first* quarter that distills over, as with this come any gases the water may have dissolved. But there are not enough of these gases in ordinary water to vitiate the result of any experiment in this book.

Store your distilled water in a clean, well-stoppered bottle.

When the laboratory is heated by steam it is advisable to condense steam from the heating apparatus and to use distilled water in many other experiments besides those in which its use is imperative.

A coil of small-bore tin pipe set in a water jacket of sheet copper makes a good condenser. The tin pipe should be connected at its upper end with the steam pipes, and its lower end should project from the copper jacket. The water jacket should be so arranged that a stream of cold water may be let in at the bottom and the warm water may flow out at the top. There should be a stop-cock, or valve, for the steam pipe, and one for the water supply.

As solid impurities are often carried into the condenser from the steam pipes, it is best to let the distilled water trickle through a large funnel containing an ordinary filter paper. Collect the water in jugs or large bottles.

Students should keep their wash-bottles filled with distilled water when the supply is abundant.

APPENDIX P.

Directions for a Student who has no Instructor.

It is necessary, of course, to have a working room that you can call your laboratory. It is not essential that this room be large and built specially for your work. Almost any room will do, provided it can be fitted with a table, a sink, and gas or gasolene.[1]

Do not think that it is necessary to have a fully-equipped laboratory before you can commence to study chemistry. Get enough apparatus to start with, and begin your work. You will soon see what you need, and how best to supply your wants. If you get into difficulty, try to *think* your way out of it. Never for-

---

[1] If it is necessary for you to use gasolene, you should be sure that you state this fact to your dealer in apparatus, and have burners suitable for this fuel sent you. If you have neither gas nor gasolene, get the largest and best alcohol lamp obtainable; but even with the best alcohol burners you may have to omit a number of experiments. Before resorting to the use of alcohol, you should do your best to obtain gas or gasolene.

get that there are more ways than one for attaining your end.

Your room best have a supply of water that can be turned on and off by means of a faucet at the sink. If you do not have town or city water delivered under pressure, the best way is to arrange a small tank above your sink. The higher this tank is above the sink the better, although two or three feet will do for almost all work. A wooden box serves well for the tank. Get a plumber to line it with zinc or copper, and fit to it a piece of pipe [of any kind], terminating with a faucet over your sink. If the plumber can arrange a pump to deliver water directly into your tank, so much the better. Otherwise you must fill the tank by pails. It is well to have a second faucet with a small delivery tube over which you can slip a piece of quarter-inch rubber tube. This second faucet should be a foot or more above the bottom of the sink, and is needed in a few cases, *e.g.*, when the filter pump is required. This second faucet, however, is not absolutely necessary, as a good substitute may be obtained by inserting a one-hole stopper in the opening of the other faucet, and fitting a short piece of glass tube to the faucet by means of the cork. Over this glass tube can be slipped the rubber tube.

The arrangement for your gas supply is most simple. All that is needed is a gas cock terminating in a corrugated nozzle over which may be slipped the quarter-inch rubber tube of the Bunsen burner. "Shut-offs" of this kind, ready to be screwed into a tee or an elbow, come in trade. If your plumber does not have any, get him to fit the end of your gas pipe with a common "shut-off," or cock, to the projecting end of which has been fitted a brass "pillar" taken from a common bat-wing burner. The clay tip should be removed from the brass pillar before the rubber tube is slipped on. It is well to have three of these delivery places for gas, although almost all experiments can be performed with a single gas tap. The blast-lamp does not need any larger supply of gas than that furnished by one of these taps. The three taps may be all close together or at a distance from one another, as the gas can easily be conveyed in rubber tubes wherever needed. It is, however, best to have the taps all so near the work table that, when standing at your work, you can reach them and turn the gas on or off. A special gas fitting is not absolutely essential, as the rubber tube for the burner may be slipped on any ordinary gas fixture.

Though almost any kind of sink will do, the best form seems to be a small iron one about twenty inches long, twelve inches wide, and five or six inches deep. The outlet to the sink should be at one end, and best be reamed out to fit an overflow plug that may be inserted in order to convert the sink into a pneumatic trough. The overflow plug is simply a piece of pipe open at both ends, an inch or so shorter than the depth of the sink, and tapered to fit the reaming of the outlet to the sink. When the plug is put in the hole and the faucet opened, the water can rise in the sink no higher than the top of the plug, for at the top it finds an outlet down the plug into the drain. As a bridge for this pneumatic trough, provide a piece of stout galvanized sheet-iron about two and a half inches wide and twelve inches long. In the middle of this strip of iron make a circular hole about half an inch in diameter. Then, at points an inch and a half from each end of the strip, bend the strip at right angles and form a kind of platform, thus: ⌐⌐ on which bottles, jars, *etc.*, can be placed, in order to collect gases when this iron bridge is set on the bottom of the sink. If you cannot use your sink for a pneumatic trough, order a small trough from the apparatus dealer, or make use of a deep pan or, better, earthen dish.

You should also have some shelves and a drawer or two for storing apparatus and chemicals.

There should be an earthen slop-jar for waste material.

Keep your table neat. Do not let dirty dishes collect. Have a clearing off of apparatus at the end of every day's work, just as if you were a member of a class in a laboratory where strict rules for neatness and order are enforced.

INDEX.

BOOK X

INDEX.

A.

Absolute scale, 145.
Acids,
  carbonic, 33.
    with lime water, 47.
    with sodium hydroxide, 54.
  hydriodic, 80.
  hydrobromic, 78.
  hydrochloric,
    preparation, 56, 57.
    solubility, 58.
    with ammonium hydroxide, 71, 108, 131.
    with marble, 58.
    with sodium, 59.
    with sodium hydroxide, 60.
  hydrofluoric, 81–82.
  muriatic, 58.
  nitric, 64.
    with ammonium hydroxide, 72.
    with carbon, 66.
    with copper, 65.
    with magnesium, 65.
    with potassium hydroxide, 67.
  phosphoric, 33.
  sulphuric, 29.
    removal of hydrogen from, 30.
    with ammonium hydroxide, 72.
    with calcium hydroxide, 46.

Acids — sulphuric,
    with iron sulphide, 40.
    with magnesium, 43.
    with marble, 51.
    with potassium hydroxide, 62.
    with sodium carbonate, 55.
    with sodium hydroxide, 54.
    with water, 30, 106.
    with zinc, 37.
    with zinc oxide, 38.
  sulphurous, 27.
Action of acids,
  on aluminum, 99.
  on gold, 96.
  on lead, 91.
  on platinum, 98.
  on silver, 93.
  on tin, 90.
Agricola, 122.
Aids for determining atomic weights,
  I. Law of Gay-Lussac and hypothesis of Avogadro, 192.
  II. Law of Dulong and Petit, 208–209.
  III. Law of Isomorphism, 211.
  IV. Periodic Law, 213.
Air,
  with hydrogen, 21–22.
  with iron, 11.
  with phosphorus, 13.

262 INDEX.

Air,
 composition of, 153.
 effect of pressure on, 126-129.
 weight and specific gravity of, 141.
Alchemy,
 period of, 114-119.
Alkaline substances, 53.
Alloy, 93.
 fusible, 93.
Alum, 100.
Aluminum, 99.
 properties, 99.
 oxidation, 99.
 with acids, 99.
 sulphate, 99.
Amalgam, 55.
 gold, 97.
 sodium, 55.
Ammonia, 68.
 fountain, 70.
Ammonium, 71.
 hydroxide, 71.
 chloride, 71.
 nitrate, 72.
 sulphate, 72.
Ampère, 180.
Analysis, 16, 107, 131.
 qualitative, 130-134.
 quantitative, 164-167.
 proximate, 107.
 ultimate, 107.
 of marble, 50-51.
 of table salt, 165.
Analytical chemistry, 107.
Anhydride, 49.
Antimony, 87.
 properties, 87.
 oxide, 87.
 chloride, 88.
 hydride, 88.

Antimony,
 solution for testing, 249.
Apparatus,
 for chemical work, xxix.
 for electrolysis of water, 239.
Aqua, 221.
 ammonia, 70.
 regia, 97.
Aqueous vapor,
 effect on gas volume, 174.
Arabs, 115.
Aristotle, 112.
 theory of the elements, 112-113.
Arsenic, 83.
 properties, 83.
 oxide, 83.
 reduction of the oxide, 84.
 hydride, 84.
 mirror, 84, 85.
 detection of, 85.
 solution for testing, 249.
Arsenide of hydrogen, 84.
Arseniuretted hydrogen, 84.
Arsine, 84.
Assaying, 131.
Atom, 170.
Atomic
 theory, 170.
 weights, 171, 177.
 table of, 216.
Aurum, 96.
Avogadro, 180.
 his hypothesis, 180.

Barium, 134.
 nitrate, 159.
Becher, 138.
Berthollet, 162.
Bismuth, 89.
 properties, 89.

INDEX.

Bismuth,
  nitrate, 89.
  basic nitrate, 89.
Black, 140.
Blank-books, xxiii.
Blast-lamp, 236.
Blow-pipe, 250.
Boyle, 125.
  law of, 126.
  tests of, 132.
  chemical theory of, 135.
  period of, 125–137.
Bromides,
  hydrogen, 78.
  sodium, 78.
Bromine, 77.
  properties, 77.
  replaced by chlorine, 78.
  will replace iodine, 81.
Buddha, 113.
Burner,
  alcohol, 254.
  gasolene, 254.
  Bunsen, 235.
  Bunsen blast, 236.

C.

Calcium, 44.
  properties, 44.
  with water, 45.
  oxide, 44.
    with water, 45.
  hydroxide, 45.
    with carbonic acid, 47.
    with sulphuric acid, 46.
  hydrate [see hydroxide].
  carbonate, 47.
  chloride, 59.
  fluoride, 81.
  sulphate, 46, 52.
  light, 50.

Carbon, 14.
  properties, 14.
  with oxygen, 18.
  oxides,
    non-combustible or dioxide, 19, 35.
      with calcium hydroxide, 49.
      with potassium, 61.
      with sodium hydroxide, 55.
      with water, 33.
      with zinc, 35.
    combustible or monoxide, 35–36.
      with red oxide of mercury, 37.
  action with nitric acid, 66.
  sulphide, 24.
Carbonates,
  calcium, 47.
  sodium, 54.
    with sulphuric acid, 55.
Carbonic acid, 33.
  with calcium hydroxide, 47–49.
  with sodium hydroxide, 54.
Calorie, 204.
Calorimeter, 204.
Catch-bottle, 244.
Cavendish, 141.
Chalk, 47.
Changes, 103.
  chemical, 103–104.
  physical, 103.
  analytical, 107.
  metathetical, 109.
  synthetical, 108.
  caused by water, 105–103.
Charcoal, 14.
Charles,
  law of, 143.
Chemeia, 115.
Chemi, 115.

264 INDEX.

Chemical,
  changes, 103–104.
  symbols, 177, 218.
  formulae, 219.
  equations, 221–223.
  Examination, 89.
  Investigation, 63.
Chemicals, xxix.
Chinese,
  early chemical knowledge of, 111.
Chlorate of potassium, 16, 25.
  molecular weight of, 198.
Chlorides, 56.
  ammonium, 71.
  antimony, 88.
  calcium, 59.
  gold, 97.
  hydrogen, 56.
  lead, 92.
  potassium, 199, 212.
  sodium, 57.
Chlorine, 56.
  preparation, 247.
  properties, 56.
  nascent, 97.
  replaces bromine, 78.
  replaces iodine, 80.
Combining number, 172.
  for zinc, 172.
Combustion products,
  of a candle, 157.
Compound, 12.
Conservation of mass, 157.
  law of, 157.
Constant weight, 26.
Contents,
  table of, xiii.
Cooke, 190.
Copper, 41.
  properties, 41.

Copper,
  oxide, 41.
    reduction of the oxide, 42.
  with nitric acid, 65.
  replaces silver, 94, 95.
Corks,
  to render tight, 230.
Crith, 178.
Crystallization,
  water of, 31, 106.
Crystal,
  to nurse a crystal, 251.

D.

Dalton's
  law, 143–145.
  atomic theory, 170.
Death of a metal, 117.
Decant, 47.
Definite proportions by volume,
  law of, 179.
Definite proportions by weight,
  law of, 163.
Deliquescence, 53.
Density, 142.
Diagrams,
  to be drawn, 32, 56.
Diamond, 14, 156.
Displacement,
  catch by, 25.
Drying,
  of bottles, *etc.*, 234.
Dulong and Petit, 202.

E.

Earliest period, 111–113.
Efflorescence, 54.
Egypt, 115.
Egyptians,
  early chemical knowledge of, 111.

INDEX. 265

Electrolysis,
  of water, 19–20, 180.
Elements,
  list of common, 134.
  list of all recognized, 219.
Empedocles, 113.
English system of weights and measures, 5.
Equations,
  chemical, 221–222.
Etching of glass,
  by hydrofluoric acid, 82.
Examination,
  a chemical, 89.
Expansion,
  irregular of liquids, 184.
  regular of gases, 185.
Explosion,
  air and hydrogen, 21, 241.
  tube, 42, 242.

F.

Factor, 38.
Ferric chloride, 110.
Filter,
  paper, 233.
  pump, 243.
Filtrate, 43.
Fluorides, 81.
  calcium, 81.
  hydrogen, 81.
    etching of glass, 82.
Fluorine, 81.
Formulae, 219.
Fusible alloy, 93.

G.

Gas,
  origin of the term, 121.
  illuminating,
    weight and specific gravity of, 149.

Gas,
  sylvestre, 121, 140.
Gas-retort carbon, 14.
Gay-Lussac, 179.
  law of, 179.
Geber, 118.
  sulphur-mercury theory, 118.
Generator, 246.
Glass,
  to bend, 228.
  to bore, 231.
  to cut, 226.
  to draw, 229.
  to fire-polish, 228.
  to mark, 226.
  to pass through rubber, 231.
Glauber, 123.
Glauber's salt, 123.
Gold, 96.
  properties, 96.
  color, 97.
  with acids, 96.
  chloride, 97.
  amalgam, 97.
Graphite, 14.
Gypsum, 46.

H.

Halogen, 81.
Hard water,
  temporarily, 48.
  permanently, 48.
Heat, 203.
  specific, 202.
    of iron, 207.
    of zinc, 205.
    table, 208.
Historical periods,
  earliest, 111–113.
  of Alchemy, 114–119.

Historical periods,
  medical, 120–124.
  of Boyle, 125–137.
  of phlogiston, 138–139.
  pneumatic, 140–160.
  modern, 161–217.
Holder,
  for test-tubes, *etc.*, 232.
Hood, 247.
Hydrates [see hydroxides].
Hydration, 106.
Hydriodic acid, 80.
Hydrobromic acid, 78.
Hydrofluoric acid, 81.
Hydrochloric acid, 56.
  preparation, 56, 57.
  properties, 58.
  solubility, 58.
  with marble, 58.
  with sodium, 59.
  with sodium hydroxide, 60.
Hydrogen,
  preparation, 20, 30, 37.
  properties, 20–23.
  lightness, 22.
  weight and specific gravity, 148.
  with air, 22.
  explosions, 21, 241.
  atomic weight, 178, 193.
  molecular weight, 193.
  removed from sulphuric acid, 30, 37.
  arsenide, 84.
  antimonide, 88.
  bromide, 78.
  chloride, 56.
  fluoride, 81.
  sulphide, 39, 40.
  sulphate [sulphuric acid].
  sulphuretted, 41.
  arseniuretted, 84.

Hydroxides,
  ammonium, 71.
  calcium, 45.
  potassium, 61.
  sodium, 53.

I.

Illuminating gas,
  weight and specific gravity of, 149.
Indestructibility of matter, 157.
Investigation,
  a chemical, 63.
Iodides,
  potassium, 80.
  sodium, 211.
Iodine, 79.
  properties, 79.
  solubility, 79.
  tincture, 79.
  action on the skin, 79.
  action on starch, 79.
  replaced, 80, 81.
Iron, 11.
  properties, 11.
  with air, 11.
  with oxygen, 17.
  oxide, 12, 18.
    with water, 32.
  chloride, 110.
  sulphate, 31.
  sulphide, 39, 135.
Isomorphism, 211.
  discovered, 211.
  applied, 211–212.

J.

Jews,
  early chemical knowledge of, 111.
Joints,
  to render tight, 231.

K

Kalium, 60.
King of metals, 96.
Kjeldahl flask, 16.

L

Language,
  of chemistry, 218.
Laughing gas, 72.
Lavoisier, 153.
Law of,
  Boyle, 126.
  Charles [see Dalton's].
  conservation of mass, 157.
  Dalton, 143–145.
  definite proportions,
    by volume, 179.
    by weight, 163.
  Dulong and Petit, 202.
  Gay-Lussac, 179.
  isomorphism, 211.
  multiple proportions, 170.
  periods, 214.
  specific heats, 208–209.
Lead, 91.
  properties, 91.
  with acids, 91.
  oxide, 91.
    with water, 91.
  chloride, 92.
  sulphate, 92.
  tree, 92.
  replaced by zinc, 92.
Libavius,
  first chemical text-book, 120.
Lime,
  light, 50.
  quick, 44.
  slacked or slaked, 45.
  water, 47.
Litmus papers, 27, 242.

M

Magnesium, 43.
  properties, 43.
  with nitric acid, 65.
  with sulphuric acid, 43.
  oxide, 43.
    with water, 43.
  sulphate, 43.
Manganese, 26.
  black oxide, 25.
Marble, 47.
  analysis of, 50–51.
  with sulphuric acid, 51.
  with hydrochloric acid, 58.
Mass,
  conservation of, 157.
Matrass, 15, 230.
Matter,
  indestructibility of, 157.
Mayow, 155.
Measuring, 3.
Medical period, 120–124.
Mendeléeff, 213.
Mercury, 14.
  properties, 14.
  with gold, 97.
  with sodium, 55.
  oxide, 15, 37, 153.
Metal,
  death of, 117.
  resurrection of, 117.
Metals,
  with platinum, 98.
  noble, 97.
Metathesis, 109, 110.
Metric system,
  of measures and weights, 3, 5.
Meyer, 213.
Microcrith, 178.
Millicalorie, 204.

Mirror,
  arsenic, 84, 85.
Mitscherlich, 211.
Mixture,
  and chemical compound, 135.
Modern period, 161-217.
Molecular theory, 180.
Molecular weight, 192.
  determined by Chemical Method, 197.
  determined by Physical Method, 196.
  of carbon dioxide, 196.
  of chlorate of potassium, 198.
  of chloride of potassium, 199.
  of hydrochloric acid, 201.
  of hydrogen, 193.
  of oxygen, 194, 196.
  of sulphate of potassium, 200.
  of sulphuric acid, 200.
  of water, 194.
Molecules, 176.
  their size, 189-190.
  their movements, 188-189.
Multiple Proportions,
  law of, 170.
Muriatic acid, 58.

N.

Nascent chlorine, 97.
Natrium, 52.
Neutral, 54.
Neutralization, 54.
Newlands, 213.
Nickel, 134.
Nitrates,
  ammonium, 72.
  bismuth, 89.
  potassium, 67.
Nitre, 63.

Nitric acid, 64.
  with ammonium hydroxide, 72.
  with carbon, 66.
  with copper, 65.
  with magnesium, 65.
  with potassium hydroxide, 67.
Nitric oxide, 72.
Nitrogen, 63.
  discovered, 150.
  properties, 63.
  oxides, 72.
Nitrous oxide, 72.
Noble metals, 97.
Note-book, xxiii.
  method of keeping, xxiv.
N. T. P., 146.
Nursing a crystal, 251.

O.

Oxides,
  arsenic, 83.
    reduction of, 84.
  antimony, 87.
  calcium, 44.
    with water, 45.
  carbon,
    monoxide, 35-36.
      oxidation of, 37.
    dioxide, 19, 35.
      with lime water, 49.
      with potassium, 61.
      with sodium hydroxide, 55.
      with zinc, 35.
      with water, 33.
  copper, 41.
    reduced by hydrogen, 42.
  hydrogen, 22.
  iron, 12, 18.
  lead, 91.
    with water, 91.

INDEX.

Oxides,
  magnesium, 43.
    with water, 43.
  manganese, 25.
  mercury, 15, 37, 153.
  nitrogen, 72.
  phosphorus, 14.
    with water, 33.
  potassium, 61.
    with water, 61.
  silver, 95.
  sodium, 52.
    with water, 53.
  sulphur, 27.
    gaseous, 27.
      with water, 27.
    solid, 27–29.
      with water, 29–30.
  tin, 90.
  zinc, 34.
    with water, 34.
    with sulphuric acid, 38.
Oxidation, 19, 37.
Oxygen, 12.
  discovery, 153.
  preparation, 15, 16, 19, 25.
  properties, 15–20.
  with iron, 12, 17.
  with phosphorus, 13, 18.
  with carbon, 18.
  with sulphur, 25–27.
  atomic weight, 172, 194.
  molecular weight, 194, 196.

P.

Palissy, 122.
Paracelsus, 120.
Paris,
  plaster of, 46.
Periodic law, 214.
Petit and Dulong, 202.

Phenomenon, 12.
Philosophers' stone, 118.
Phlogiston, 138.
  period of, 138–139.
Phoenicians,
  early chemical knowledge of, 111.
Phosphoric acid, 33.
Phosphorus, 12.
  properties, 12.
  with air, 13.
  with oxygen, 18.
  oxide, 14.
Pipette, 48.
Plaster of Paris, 46.
Platinum, 98.
  properties, 98.
  with acids, 98.
  with metals, 98.
  with other chemicals, 98.
  sponge, 28, 98.
Plumbers' solder, 92.
Plumbum, 91.
Pneumatic period, 140–160.
Potassium, 60.
  properties, 61.
  with air, 61.
  with water, 61.
  with dioxide of carbon, 61.
  oxide, 61.
    with water, 61.
  chlorate, 16, 25, 198.
  chloride, 199, 212.
  hydroxide, 61.
    with water, 61.
    with nitric acid, 67.
  iodide, 80.
  nitrate, 67.
  sulphate, 62.
Precipitate, 43.
  to dry, 251.

INDEX.

Pressure,
    effect on air, 126–129.
Priestly, 150.
Product, 38.
Proportions,
    multiple, law of, 170.
    by volume, law of, 179–180.
    by weight, law of, 163.
Proust, 162, 168.
Prout, 175.
Proximate analysis, 107.
Pump,
    air, 142.
    filter, 243–244.

Q.

Qualitative analysis, 131.
Qualitative tests, 132.
Quantitative analysis, 164.
Quick lime, 44.

R.

Reaction, 43.
Reduction, 36.
    of oxide of carbon, 35–36.
    of oxide of copper, 42.
    of oxide of mercury, 15, 37.
    of chloride of silver, 96.
Resurrection,
    of a metal, 117.
Richards, 216.
Richter, 161.
Rubber,
    to cut, 231.
Rutherford, 150.

S.

Sal soda, 54.
Salt,
    definition, 67.
    table, 57.
    made, 57, 60.
Salt,
    analyzed, 165.
Scheele, 150.
Scheele's green, 152.
Silver, 93.
    properties, 93.
    oxidation, 93, 95.
    oxide, 95.
    with acids, 93.
    chloride, 94, 96.
    sulphide, 95.
    replaced by copper, 95.
    purification, 95.
Simple substance, 12.
Slaked lime, 45.
Solder, 92.
Soot, 14.
Sodium, 52.
    properties, 52.
    with air, 52.
    with water, 53.
    oxide, 52.
        with water, 53.
    amalgam, 55.
        to prepare on a large scale, 248.
    bromide, 78.
    carbonate, 54.
        with sulphuric acid, 55.
    chloride, 57.
    hydrate [see hydroxide].
    hydroxide, 53.
        with carbonic acid, 54.
        with carbonic dioxide, 55.
        with hydrochloric acid, 60.
        with sulphuric acid, 54.
    with hydrochloric acid, 59.
Solution, 103–106.
Spaces,
    between the molecules, 182.
Spain, 115.

INDEX. 271

Specific gravity, 142.
  of air, 141.
  of carbonic dioxide, 145.
  of hydrogen, 148.
  of illuminating gas, 149.
  of oxygen, 196.
Specific heat, 202.
  of iron, 207.
  of zinc, 205.
  table, 208.
Spencer, 171.
Sponge,
  platinum, 28, 98.
Stahl, 138.
Stalactite, 48.
Stalagmite, 48.
Stannum, 89.
Starch,
  with iodine, 79.
Stibium, 87.
Stoichiometry, 224.
Stoppers,
  to prevent mixing, 232.
  to remove when stuck, 234.
Sublimation, 72.
Substances,
  simple, 12.
  compound, 12.
Suidas, 115.
Sulphates,
  aluminum, 99.
  ammonium, 72.
  calcium, 46, 51.
  iron, 31.
  lead, 92.
  magnesium, 43.
  potassium, 62.
  sodium, 55.
  zinc, 38.
Sulphides,
  carbon, 24.

Sulphides,
  hydrogen, 39, 40.
  iron, 39, 135.
  silver, 95.
Sulphur, 23.
  properties, 23.
  modifications, 24.
  oxides, 25, 27.
  with hydrogen, 39.
  with iron, 39, 135.
  with zinc, 136.
Sulphuretted hydrogen, 41.
Sulphuric acid, 29.
  removal of hydrogen from, 30, 37.
  with ammonium hydroxide, 72.
  with calcium hydroxide, 46.
  with iron sulphide, 40.
  with magnesium, 43.
  with marble, 51.
  with potassium hydroxide, 62.
  with sodium carbonate, 55.
  with sodium hydroxide, 54.
  with water, 30, 106.
  with zinc, 37.
  with zinc oxide, 38.
Sulphurous acid, 27.
Symbols,
  Dalton's, 177.
  modern chemical, 218.
Synthesis, 50, 108.

T.

Temperature, 202.
Terra pinguis, 138.
Test papers, 242.
Tests,
  qualitative, 132.
  used by Boyle, 132.
  by chemical changes, 133.
  by physical changes, 132.

INDEX.

Time,
  required in the laboratory, xxvii.
Tin, 89.
  properties, 89.
  cry, 90.
  crystalline structure, 90.
  oxide, 90.
  with acids, 90.
  salts of, 90.
  replaced by zinc, 91.
  plate, 90.
Tincture,
  of iodine, 79.
Transference of motion, 203.
Transmutation, 114, 116.
Turmeric paper, 242–243.

U.

Ultimate analysis, 107.

V.

Valentine, 119.
Van Helmont, 120.

W.

Wash-bottle, 6.
Water,
  preparation, 22.
  electrolysis, 19–20, 180.
  with iron oxide, 32.
  with phosphorus oxide, 33.
  with carbonic dioxide, 33.

Water,
  with gaseous oxide of sulphur, 27.
  with the second oxide of sulphur, 29–30.
  with zinc oxide, 34.
  with magnesium oxide, 43.
  with calcium oxide, 45.
  with sodium oxide, 53.
  with potassium oxide, 61.
  with lead oxide, 91.
  with sodium, 53.
  with potassium, 61.
  with sulphuric acid, 30, 106.
  of crystallization, 31, 106.
  permanently hard, 48.
  temporarily hard, 48.
  distilled, 252.
  molecular weight, 194.
Water-bath, 26, 185.
Weighing, 4.
Weight,
  constant, 26.

Z.

Zinc, 34.
  properties, 34.
  oxide, 34.
    with sulphuric acid, 38.
  replaces tin, 91.
  replaces lead, 92.
  combining number for, 172.
  sulphate, 38.
  with carbonic dioxide, 35.

www.ingramcontent.com/pod-product-compliance
Lightning Source LLC
Chambersburg PA
CBHW022112230426
43672CB00008B/1359